CTI Solutions and Systems

Other McGraw-Hill Books of Interest:

Bates/Gregory * Voice/Data Communications Handbook * 0-07-005147-X

Bezar * LAN Times Guide to Telephony * 0-07-882126-6

Goralski * Intro to ATM Networking * 0-07-024043-4

Grinberg * Computer/Telecom Integration: the SCAI Solution * 0-07-024842-7

Heldman * Future Telecommunications * 0-07-028039-8

Hodge * Interactive Television: A Comprehensive Guide for Multimedia Technologists * 0-07-029151-9

Keyes * The McGraw-Hill Multimedia Handbook * 0-07-034475-2

Kumar * Broadband Communications * 0-07-035968-7

Lee * Mobile Cellular Telecommunications Systems * 0-07-038089-9

Lindberg * Digital Broadband Networks and Services * 0-07-037936-X

McDysan/Spohn * ATM: Theory and Application * 0-07-060362-6

Minoli * ATM & Cell Relay Services in Corporate Environments * 0-07-042591-4

Minoli * Video Dialtone Technology * 0-07-042724-0

Pecar * McGraw-Hill Telecommunications Factbook * 0-07-049183-6

Russell * Signaling System 7 * 0-07-054991-5

Spohn * Data Network Design 2E * 0-07-060363-4

Summers * ISDN Implementer's Guide * 0-07-069416-8

Travis * Enterprise Telecommunication: Fundamentals for Future Networking * 0-07-57695-5

CTI Solutions and Systems

How to Put Computer Telephony Integration to Work

Michael Bayer

McGraw-Hill

New York   San Francisco   Washington, D.C.   Auckland   Bogotá
Caracas   Lisbon   London   Madrid   Mexico City   Milan
Montreal   New Delhi   San Juan   Singapore
Sydney   Tokyo   Toronto

Library of Congress Cataloging-in-Publication Data
Bayer, Michael Thomas
    CTI solutions and systems : how to put computer telephony integration to work /
    Michael Thomas Bayer
        p.   cm.
    Includes index.
    ISBN 0-07-006153-X (hc)
    1. Telephone systems—Automation.   2. Telephone systems—Data processing.   I. Title.
TK6397.B33 1997
621.385--dc21   96-29858
                                                                                CIP

McGraw-Hill

A Division of The McGraw·Hill Companies

1 2 3 4 5 6 7 8 9 0   DOC/DOC   9 0 2 1 0 9 8 7

ISBN 0-07-006153-X

*The sponsoring editor for this book was Steve Chapman, the editing supervisor was Sally Glover, and the
production supervisor was Clare Stanley. It was set in Palatino by Michael Bayer of Computer Telephony
Solutions.*

Printed and bound by R.R. Donnelley & Sons Company.

This book is printed on recycled, acid-free paper containing a minimum of
50% recycled, de-inked fiber.

McGraw-Hill books are available at special quantity discounts to use as premiums and sales promotions, or for use in corporate training programs. For more information, please write to the
Director of Special Sales, McGraw-Hill, 11 West 19th Street, New York, NY 10011. Or contact your
local bookstore.

To my parents, whose love, support, and encouragement
– which I have always taken for granted –
has been more than anyone could ever hope for.

Notices

CallPath is a trademark of International Business Machines

IEEE is a trademark of the Institute of Electrical and Electronics Engineers

GeoPort is a trademark of Apple Computer, Inc.

IrDA is a trademark of the Infrared Data Association

Mac OS is a trademark of Apple Computer, Inc.

Windows is a trademark of Microsoft, Inc.

USB is a trademark of Intel Corporation

Versit is a trademark of Apple Computer, Inc., AT&T, International Business Machines, and Siemens Rolm Communications, Inc.

All other product names and brand names are the trademarks, registered trademarks, or service marks of their respective owners.

Contents

Contents

Contents

Contents

Contents

Contents

Contents

Contents

Contents

Contents

Contents

Contents

Sidebars

Figures

1. What Is CTI?

2. CTI Solutions and Benefits

3. Telephony Concepts

Figures

4. Core Telephony Features and Services

Figures

5. Telephony Equipment and Network Services

Figures

6. CTI Concepts

7. CTI System Configurations

Figures

8. CTI Software Components

9. CTI Solution Examples

Tables

Acknowledgments

Thank you to Mark Hume, Bob Papp, Frederic Miserey, Mike Spieser, Jean Michel Karr, John Stossel, Susan Schuman, and all the other members of the extended MTA team that got this ball rolling.

Thanks to my friends Susan King, Bob Flisik, and Lisa Heller, who planted the seeds for this book.

Thank you to Dave Anderson, Frederic Artru, Marc Fath, Jose Garcia, Jim Knight, Steve Rummel, and Tom Miller for their years of dedication to making CTI Plug & Play a reality.

A big thank-you to Dave Anderson, Frederic Artru, April Bayer, Marc Fath, Lisa Heller, Tim Schweitzer, Mike Spieser, and Tamara Wilkerson for all the constructive feedback and editorial input.

Last but not least, thanks to Steve Chapman for sponsoring this project and fighting all the necessary battles.

Introduction

This book is for anyone who wants to know about Computer Telephony Integration (CTI), specifically anyone who is:

- Interested in putting CTI to work in the home or business; or

- Interested in opportunities to get involved in developing CTI solutions or products.

This book will show you how to put your time and resources to better use if you:

- Use a personal computer

- Own or manage a business

- Work from home

- Spend a lot of time on the telephone

- Manage or work in a telecommunications department

- Manage or work in an information systems (IS) department

- Manage or work in a call center or help desk

- Write application, utility, or system software

Introduction

As CTI becomes a mainstream technology, you'll find this book indispensable if you work for a:

- Telephone company

- Cable company

- PBX company

- Consumer electronics company

- Computer or electronics store

- System integrator

This book does not assume that you have any prior knowledge of telephony. All the telephony concepts and terminology you need in order to master CTI are explained and illustrated.

If you do have a telephony background, you will find this book's review of telephony from a universal perspective to be invaluable. In the world of telephony technology, every company uses a slightly different vocabulary, and this complicates CTI product and solution development. To address the requirement for a unifying model, this book presents a common set of concepts and terminology which reflect industry standards. These are explained in a fashion that makes them easy to relate to the models with which you are familiar.

As a reader of this book, you should have some level of familiarity with using and operating your favorite personal computer or information appliance (e.g., Windows PC, Macintosh, OS/2 PC, AIX workstation, Novell server, Newton PDA, etc.).

The Time for CTI Is Now!

The field of CTI technology has been developing slowly but steadily for more than a decade. Early pioneers and visionaries (the author among them) have championed and struggled to bring about the ubiquitous integration of computers with telephony functionality for longer than the abbreviation "CTI" has existed.

Unfortunately, CTI is necessarily a complicated topic. The results of more than 100 years of frenetic innovation in the telephony world, and an even faster and more innovative explosion of technology in the computer world, meant that considerable effort was required to build a framework for no-compromise CTI solutions. The bulk of the work has now been done, however, and this book represents the result.

Despite the challenges involved, the emergence of CTI as an essential element of everyone's technology environment is inevitable. Many incremental developments in the telephony, telecommunications, and computer industries have now brought CTI to the point where it has emerged as a mainstream technology.

Three key factors mark the maturation of CTI technology:

- *Interoperability*, through standard CTI Plug and Play protocols

- *Power*, though easy-to-use software tools

- *Access*, through low-cost multimedia communications hardware

This book provides a thorough introduction to telephony and CTI technology based on the latest industry standards. These represent a universal abstraction, or unification, of the historically divergent telephony world. The framework presented by this book will help you understand the capabilities, limitations, and potential of existing products and equipment, and to anticipate the next generation (and future generations) of new telephony and CTI products. You will be able to make the right CTI planning and implementation decisions now, in anticipation of future CTI developments, and you won't get caught heading down a technological dead end.

Organization of this Book

To be a passenger on an airplane you really don't need to know anything about the physics of flying, the training and licensing of pilots, the safety record of different airlines, or even how to buy a ticket. Whether you are an airplane designer, a new pilot, an investor in airline stocks, or just a passenger, however, a little bit of insight into each of these areas can greatly enhance your enjoyment of the experience and the likelihood of your success.

With this in mind, the goal of this book is to provide you with a complete insight into CTI and its underpinnings, regardless of what your final objective might be. In fact, the more you learn about the endless possibilities for taking advantage of telephony and CTI, the more you'll want to put it into practice!

Chapter 1: What Is CTI?

Chapter 1 describes exactly what CTI is and is not. It provides context for the discussion of CTI and specifically relates CTI technology to the broader category of CTI solutions. This chapter presents the fundamental insights into why CTI is now maturing, and why it is likely to become as ubiquitous as the graphical user interface and other mainstream personal computing technologies.

Chapter 2: CTI Solutions and Benefits

Chapter 2 presents a collection of eight complete CTI solution scenarios in which people from different walks of life have applied CTI technology to improve their working and living environments. This chapter explains the CTI value chain, which reflects all of the different services and components that must be integrated to build functional CTI solutions. The CTI value chain provides the hierarchy around which the rest of the book is structured.

Chapter 3: Telephony Concepts

Chapter 3 presents a thorough introduction to the telephony concepts you need to know in order to take full advantage of CTI technology. This chapter presents a complete and universal abstraction of the world of telephone system technology. This simple abstraction is used throughout the book as a framework for easily describing any telephony product or capability.

Chapter 4: Core Telephony Features and Services

Chapter 4 builds on the concepts presented in Chapter 3 and shows how the elements within telephone systems of any size interact to provide the comprehensive range of telephony features and services commonly available.

Chapter 5: Telephony Equipment and Network Services

Chapter 5 relates the concepts presented in Chapter 3 and Chapter 4 to tangible telephony product and service implementations. It describes the options that exist for those assembling a telephone system ranging in size from a single telephone to a private telephone network.

Chapter 6: CTI Concepts

Chapter 6 introduces the concepts of CTI as a means for observing and controlling telephone systems of any size. This chapter discusses how telephony features and services are accessed through a CTI interface, and presents the capabilities of a telephone system that are specific to CTI interfaces.

Chapter 7: CTI System Configurations

Chapter 7 presents the diversity of physical CTI configurations that are now possible and that will become easier and easier to assemble, given the maturation of CTI technology and availability of products based on standardized CTI protocols.

Chapter 8: CTI Software

Chapter 8 complements Chapter 7 by presenting the diversity of software configurations that are available to support application development, CTI system integration, and user customization of CTI solutions.

Chapter 9: CTI Solution Examples

Chapter 9 revisits the CTI solution scenarios presented in Chapter 2 and looks under the covers to explain how each can be implemented using the technologies explored in the book.

Introduction

To pull everything together, each chapter ends in a review of the key concepts covered . While the book is intended to be read from beginning to end, it is designed to permit the use of individual chapters. References to key concepts explained in other parts of the book will make it simple for you to navigate.

 This special icon is used throughout the book to mark key definitions.

 This special icon is used to mark fundamental concepts of which to take special note.

 This special icon is used to mark points that clarify sources of potential confusion.

You will find that this book is extensively illustrated, to help make the frequently complex and abstract concepts surrounding telephony and CTI intuitive and easy to comprehend. Standardized graphical notations are used wherever applicable. The symbols and icons used in these graphical notations are summarized on the inside of the front and back covers for easy reference.

Sidebars and footnotes are used throughout the book to present material that is useful but nonessential. You can read these for diversion and further insight as you proceed through the book, or you can return to them later.

Exploring Further

This book is intended for users of any type or brand of telephone system, computer, or operating system. To avoid any bias in the illustrations used throughout the book, all diagrams use stylized icons to represent the various components of a system. References to particular products are made only when the product in question represents a recognized, de facto standard.

This book deals with telephony from a macro perspective: how discrete CTI-enabled telephony and computer products are integrated and assembled into CTI systems. An independent field of computer-telephony technology involves using off-the-shelf computer components to build the telephony products themselves. Beyond the fact that these computer-based telephone systems typically provide a CTI interface, this area is outside the scope of this book. For more information on this area of computer-telephony technology, you may wish to consult reference materials on the S.100 specifications from the ECTF (Enterprise Computer Telephony Forum) or its forerunners, such as SCSA.

Use this book as a resource to help you identify your CTI opportunities; determine the way that you want to approach CTI and the type of solutions, products, or components you want to assemble or build; and assess your needs and preferences. Armed with the insights you'll have, you can identify the products and the strategies that best satisfy your needs. The pace of new product development in the world of telephony and CTI is so rapid that product-specific information is not included here; it would become obsolete far too quickly. To find out about the latest in available products, you might wish to:

- Contact your existing telephone equipment vendor(s);

- Subscribe to industry magazines;

- Monitor the efforts of groups such as ECMA, ECTF, and Versit;

- Attend industry trade shows; and

- Check the author's Web site at *http://WWW.CTEXPERT.COM.*

If you are interested in developing CTI products, consult the bibliography at the back of this book for a list of the essential development documentation.

1.
What Is CTI?

Telephones and computers indisputably are the two technologies that most greatly impact every aspect of our daily lives. These technologies are central to the operation of virtually any business of any size. A vast number of organizations exist only because of these technologies. In fact, one could argue that these are the technologies that hold together the very fabric of the modern information-oriented economy. Computer Telephony Integration, or CTI, brings these two technologies together and harnesses their synergy.

The telephone network is a tremendous resource, of which everyone should be able to take full advantage. Telephony[1-1] technology, that is, the technology that lets people use the telephone network, is extremely powerful and has great potential for empowering people in every walk of life.

For the vast majority today, however, the only means of accessing the telephone network is through a telephone set or answering machine of some sort. The one thing all these devices have in common is their

---

1-1. Telephony — Incidentally, the word "telephony" is pronounced "teh•LEF•eh•nee." It should not be pronounced "teh•leh•FOH•nee" as this generally leads to some embarrassment.

arbitrary and very limited user interface. The telephone set is a terrific device for getting sound into and out of the telephone network, but it is a very poor device for getting access to powerful technology. Even the simplest features are rarely mastered by the average telephone user. All too often you've heard someone on the telephone say, "I'll try to transfer you now—but if this doesn't work, here's the number to call."

CTI changes all of this. CTI provides an alternative means of accessing the power of telephony technology. Computer technology is an empowering technology because it allows people to extend their reach, and it can take on time-consuming or non-creative tasks. Computer technology allows for tremendous customization, so that you can have a user interface and work environment that is optimized for your needs. With CTI, this means you can have full access to the power of the telephony technology you need, and have it in the form that is best for you. A computer working on your behalf can take actions independently, so it effectively becomes your assistant. With CTI technology, this means your computer can screen calls, handle routine requests for information without your intervention, and interact with callers in your absence. As shown in Figure 1-1, CTI technology not only gives you full access to telephony technology, it actually amplifies its utility!

The following examples illustrate how computer technology has the power to amplify the power of telephony and make it more accessible.

- At a pay phone:
    - *Without CTI*, you have to laboriously enter your carrier preference (if you can remember the access code), credit card information, and number you want to dial.

    - *With CTI*, to place a call you simply point your PDA or laptop computer at the IR port on the pay phone and click on a person's name. In fact, if you'll be there a while, you can have selected calls rerouted from your office number to the pay phone, which would in turn alert you with information about each incoming call.

Figure 1-1. Without CTI and with CTI

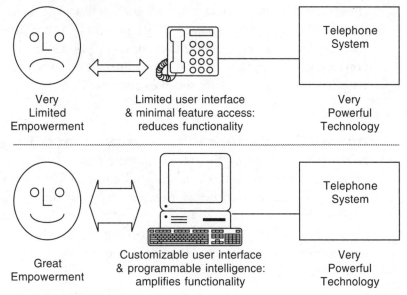

- At home:

    - *Without CTI*, you're in the living room watching TV and you remember that you have to talk to a colleague (who is traveling on the other side of the world) before he leaves his hotel. You head for the phone in your den, look up the number for his hotel, look up the country code for the country where he is traveling, and the access code. After you have dialed all of the digits correctly (on the second try) and complete your short conversation, you jot down a note to yourself as a reminder to expense the telephone call when you get your next bill. By the time you get back to the living room, you've missed the rest of your TV program.

    - *With CTI*, you wait for the next commercial break and then, with your TV's remote control, you pop up your personal directory as a picture-in-picture on your TV set and select your colleague's name. The TV then displays a set of locations, including the hotel where he is staying. You select the hotel and indicate you want to dial the number. The appropriate number is dialed automatically (using the cheapest available carrier); your TV then acts as a speaker phone, allowing you to converse with your colleague without leaving your seat. When the call completes, you return to your TV program. Information about the call is logged to your home finance package as an expensible item.

- While telecommuting:

  - *Without CTI*, you appreciate the opportunity to work at home as a means of avoiding the long commute and the other inefficiencies of work at the office. Unfortunately, however, it means you miss many important telephone calls that then wind up in your voice mail box at the office. You don't want to forward the calls home because you don't want them inadvertently being directed to your family's residential voice mail system.

  - *With CTI*, your home personal computer has a remote connection to the local area network at your office, allowing you to monitor your telephone. If an important call comes through, your computer notifies you so you can decide whether you want it to go to voice mail or be redirected to your home telephone. Using Internet telephony, you can take the call even if your phone line is busy. In fact, the system works so well that even the call center agents in your company, who spend their whole day handling customer telephone calls, are able to work from home.

- At the office:

  - *Without* CTI, you are trying to cope without the administrative support you lost through organizational downsizing. Throughout the day calls pile up in your voice mail box instead of being directed to people who could take care of callers right away. You place many telephone calls yourself, and when you need to reach someone, you spend a lot of time placing calls—only to get their voice mail systems, and you end up having to try again later on.

- *With CTI*, your desktop personal computer has taken the place of your secretary. Using CTI technology, it screens every call that comes in. If you're busy, you aren't disturbed. If the call is best handled by someone else, it is redirected without your intervention. When you need to get hold of someone, your personal computer will keep trying until the person is reached; only then does it connect you to the call, so you aren't tied up doing the redialing yourself.

- In the school:

  - *Without CTI*, there are just three telephone lines in the whole school and they can be accessed only from the principal's office and from the staff room. Despite the fact that teachers believe students would benefit greatly from the ability to use telephones as a resource for research and other projects, there are no telephones in the library or in the classrooms because there is no way of controlling their use.

  - *With CTI*, the school has an Internet telephony gateway and each of the computers in the classrooms and the library are equipped with speaker phone applications. These allow groups of children to sit around computers and place telephone calls to resource people in the community, as well as children at other schools. The school doesn't need a high-speed connection to the Internet because most calls are placed using one of the phone lines the school already has. The school's electronic bulletin board on the Internet, which lists homework assignments and lunch menus, is available not only to parents who have computers at home, but also those who want to dial in from any touchtone telephone.

- In a meeting:

  - *Without CTI*, you may miss an important call that you are expecting because you must attend an equally important meeting. You have no secretary and cannot forward all of your telephone calls to the meeting room because it would be too disruptive. You must rely on voice mail to catch the call. You leave the meeting frequently to see if it has come through and then you try to return the call.

- *With CTI*, your personal computer checks the origin of each call and identifies the important call when it arrives. Rather than simply taking voice mail, it tells your caller that it will try to track you down. It then sends a notification to the wireless PDA (personal digital assistant) you are carrying. This notification tells you who called and informs you that the caller is holding. You can respond by instructing your computer to take a message, hold the call while you return to your desk, or transfer the call to a nearby telephone. The others in the meeting are not even aware that you are having this wireless dialog with your desktop computer.

The ultimate promise of information technology is that it empowers people to collaborate with one another more effectively and with greater ease. Since the invention of the telephone over a century ago (in 1876), telephony technology has evolved tremendously. Innovation was fast and furious in the earliest days after Alexander Graham Bell's invention; it took less than two years to commercialize the technology. To this day the pace has not slowed. While modern computer technology is a much more recent invention, its history has been just as fast-paced. The key to both disciplines is their ability to improve the ways people communicate and collaborate, because these activities are at the heart of all human endeavor.

Surprisingly, the world of telephony technology and the world of computer technology until recently have remained largely isolated from one another. CTI refers to the ability to combine the products, services, and systems from these different technology areas, but it does not refer to a superset of computer and telephony technologies. CTI technology is the bridge between these technology areas (Figure 1-2). CTI permits the development of solutions that overshadow traditional applications of these individual technologies.

Figure 1-2. CTI brings together the worlds of computer and telephony technology

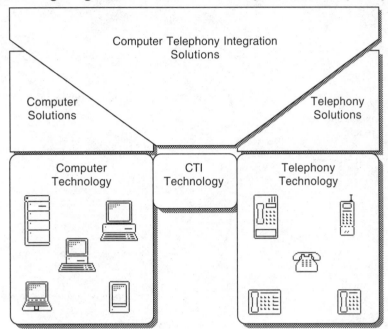

This book differentiates between "CTI technology" and "CTI solutions." These two terms are fully defined in this chapter. This book covers all the concepts and details you need to know about CTI technologies, and then illustrates how these technologies are applied in typical CTI solutions.

1.1. The Importance of Telephony

The telephone is the single most important communication appliance in the world today. Despite the high levels of acceptance for various other forms of information technology, including consumer electronics products and personal computers, the telephone (in all of its forms) remains the only communication device that can be considered ubiquitous.

Figure 1-3. Everyone you want to talk to is somewhere on the telephone network

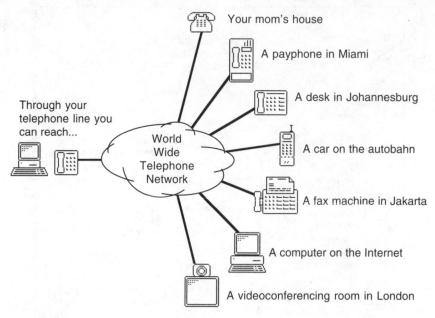

The telephone network is the single most important network in the world. This is because:

- There are more "endpoints" on this network than any other, counting just the number of connected telephones alone.

- The telephone network requires the least effort to use because its basic format is voice. Virtually any human being on the planet is able to interact with any other through the telephone network.

- The telephone network acts as both the on- and off-ramp, and as the interconnection between, most other networks in the world. In fact, there are very few people, fax machines, computers or data networks in the world that cannot be reached through the worldwide telephone network (Figure 1-3).

1.2. The Importance of Computers

The "C" in "CTI" stands for "computer." This may be a little misleading, though, because the word computer is being used here in its broadest sense. The term refers to any device that can be programmed to control or observe telephony resources. Think of it as any device using "computer technology." While this set of devices certainly includes traditional mainframes and minicomputers, as well as your own personal computer, it also includes many others. For example, *personal digital assistants* (PDAs) are small, handheld devices that generally have much less power than traditional computers and a more limited set of uses. But as personalized information appliances, PDAs are very important when it comes to telephony integration. Similarly, a whole new generation of so-called "intelligent," programmable consumer electronics products will transform everyday devices such as VCRs, TVs, watches, games, etc., into what we consider to be "computers" for purposes of CTI (Figure 1-4).

Figure 1-4. Diversity of applicable computer technologies

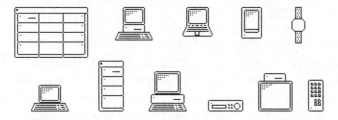

Contrary to popular wisdom, the diversity of computing devices continues to grow as computer technology become more pervasive and more specialized. In the early days of computer technology, a panel of experts is said to have concluded that only a small number of computers would ever be needed! Not so long ago it was thought that the world of personal computing revolved around one or two popular operating systems. The reality is that as innovative new devices are built for everything from automobile navigation to home entertainment, personalized computing devices become more and more diverse. Many don't even have operating systems in the traditional sense.

The motivation behind CTI is *customization* and *control*. The real significance of integrating computer technology with telephony is the opportunity to tie any form of intelligent appliance, whether it be a mainframe computer, personal computer, or other personal information appliance, into the telephone system in order to create a customized, and preferably personalized, communications environment.

1.3. Communications and Collaboration Technology

Before you conclude that CTI deals with all aspects of human communication, it's important to put CTI in the context of other areas of communications technology, or specifically, what we will refer to as *communications and collaboration* technology (C&C for short). This term refers to the superset of computing technologies and communications technologies that allow people to collaborate with one another.

1.3.1. Overall Vision for C&C

Personal computers already are being seen more as communications tools and media delivery devices than as the "document processors" they were just a few years ago. As information technology merges with communications technology, personal computers and other information appliances become the focal point for people collaborating with other people.

Information technology increasingly is taking the place of administrative staff that once provided much of the management of communications and collaboration activity. In business, secretaries used to place calls, screen calls, send faxes, type and send letters, track down people and information, take and prioritize messages, etc. Most people do not have someone to delegate all of these tasks to, instead they must perform them themselves or apply new technologies.

Collaboration Grid

People frequently described human collaboration in terms of the grid shown below. One axis represents separation by time and the other separation by distance.

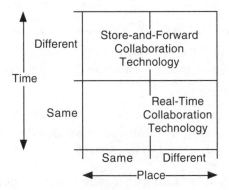

When two people are not separated by either time or distance, they can have a face-to-face conversation, read from and write on the same piece of paper, etc.

When two people are separated by time, they must use store-and-forward technology. If they are not separated by distance, this can be as simple as leaving notes on the refrigerator door (or, in ancient times, hand paintings on cave walls). It can be as technologically sophisticated as leaving a message stored in a shared computer. If the two are separated by both distance and time, then store-and-forward techniques such as postal mail, e-mail and electronic publishing come into play.

When two people are separated by distance but not by time, they rely on real-time communications technology to talk and possibly share visual information at a distance.

Vendors of communications technologies have been quick to respond to this demand and have generated a dizzying array of different and diverse communications products. While these new products and forms of communication have enriched our ability to collaborate with others, they also have added to the complexity of managing personal communications.

The personal computer and other personal information appliances offer a means for getting this complexity under control, and for being able to master fully, or put to work, the ever-richer communications infrastructure around us.

Embedding C&C technology as a central part of personal computer use involves integrating technologies from five disciplines that make up C&C. This is illustrated in Figure 1-5.

Figure 1-5. Personal computer and information appliances are becoming the focal point for people collaborating with other people.

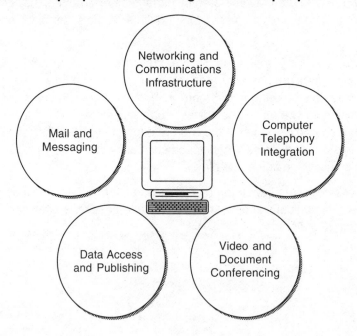

1.3.2. Five Disciplines of C&C Technology

The five disciplines, or domains, of C&C technology reflect both the key areas in which technology innovation is centered, and also the way in which people buy and use information technology. These areas can be further grouped into real-time collaboration technology, store-and-forward collaboration technology, and communications technology.

Real-time Collaboration Technology

Real-time collaboration involves technologies that overcome the barriers of distance between people who are working together at the same time.

1. Computer Telephony Integration

    Computer telephony integration (CTI) involves bringing control of the telephone and access to telephony media streams into the computer's realm.

2. Video and Document Conferencing

    Video and document conferencing complements CTI by supplementing it with shared-visual workspaces where two or more people engaged in a discussion not only have the benefit of their computer's ability to set up the call but can also take advantage of its ability to present visual information and may permit the real-time manipulation of that shared material.

Store-and-Forward Collaboration Technology

Where real-time collaboration overcomes barriers of distance between people, store-and-forward collaboration involves technologies which allow people to collaborate while working at different times, independent of location.

3. Messaging

> Messaging involves the delivery of electronic mail, fax mail, voice mail, and any other form of information that is sent by one person to one or more specified people, to be delivered at some later time.

4. Publishing and Browsing

> Publishing and browsing technologies complement messaging technologies. While messaging is a directed broadcast of information, publishing and browsing technologies allow information to be made available to a whole community of people who are granted access to the information. The Internet's World Wide Web represents the best example of this type of information publishing as a store-and-forward collaboration mechanism.

Communications Technology

The four functional areas described above each rely on network and communications "plumbing" in order to move the real-time and non-real-time data from one place to another.

5. Network and Communications Infrastructure

> Network and communications infrastructure includes hardware (such as wiring, transceivers, bridges, routers, gateways, etc.) as well as software (protocols, protocol stack implementations, drivers, etc.).

1.3.3. Bringing It All Together

The real benefit of integrating C&C technology into the fabric of everyday personal computing technology goes far beyond providing access to the various areas through a single device.

A truly complete C&C solution for a particular individual or organization draws on many select components from multiple areas to satisfy a specific set of identified needs. Such a solution derives distinct benefit from the integration that is achieved *among* the different areas.

Great products or technologies in all five areas take full advantage of other computing technologies and facilities available, from multimedia to scripting, to provide a seamless, integrated collaboration experience for a given computer user.

The potential of personal computing devices to apply intelligence, with context, and to make full use of accessible information, will take human collaboration to another level. It will make it manageable, if not effortless.

1.4. CTI: Bridging Computing and Telephony

To deliver on the vision of computing devices as the focal point for communications and collaboration, these devices must embrace the telephone. Only then can the personal computer, and other forms of information technology, fulfill their potential for making all forms of collaboration easy and accessible. While all of the areas within the C&C framework are critical in delivering on this vision, CTI is in many ways the most fundamental. CTI technology is the bridge between computing and telephony.

A quick "man on the street" poll that asked computer users what tool they use for communicating and collaborating would quickly reveal that most people identify the telephone as their most important C&C tool. If the ultimate vision for empowered human collaboration is to be achieved, CTI, or the coming together of computer technology and telephony, is on the critical path.

Migration from the Real Desktop to the Virtual Desktop

At every stage in the evolution of the personal computer, more and more of the items that once sat on our physical desktops have migrated to the virtual desktop of the personal computer.

- The typewriter was replaced by the word processor.

- The calendar was replaced by the scheduling application.

- The desktop calculator was replaced by its software counterpart.

- The dictionary was replaced by the spell checker.

- The personal stereo system was replaced by the PC's built-in CD player.

- The "In" and "Out" boxes were obsoleted by e-mail.

- The Rolodex™ was made obsolete by contact management software.

- The fax machine was replaced by scanners and fax modems.

The telephone is the only item on the desktop that hasn't been greatly affected by the information technology revolution. CTI is the technology that is bringing about the evolution of the telephone. CTI solutions at the desktop are covered in detail in Chapter 6. This is where your questions about the next step in this evolution will be answered.

A significant benefit of this function migration to the personal computer's virtual desktop is that with a laptop computer you can simply pick up your entire desk and take it anywhere you want. With the next generation of information appliances, the appropriate parts of your office will be available to you in your car, your living room, or on your refrigerator door.

1.5. Defining CTI

The notion of integrating computer technology and telephony technology has been rather amorphous up to this point. With the maturation of CTI technology, however, a very concise definition has emerged.

CTI involves three aspects:

- Call Control
    Call control is the ability to observe and control telephone calls, switching features and status, ACDs and ACD agents, and to use switching resources that include tone generators and detectors.

- Telephone Control
    Telephone control is the ability to observe and control physical telephone devices as computer peripherals.

- Media Access
    Media access involves binding telephone calls to other media services such as voice processing, fax processing, videoconferencing, and telecommunications.

CTI technology achieves functionality in each of these areas by interacting with *telephony resources* that exist within CTI-accessible telephone systems. A given CTI product may provide functionality in one, two, or all three of these areas.

1.5.1. Call Control

The most essential part of CTI is the aspect that deals with a computer observing and controlling telephone calls and the telephone system features that interact with calls. *Observing* means tracking all call processing activity that takes place, and being aware of any changes to feature settings. *Controlling* refers to issuing instructions for the telephone system to obey with respect to telephone calls, features, and associated telephony resources. Call control functionality supported through a CTI interface includes both the telephony functionality available to users of a telephone system through telephone sets, as well as functionality that applies only to computer observation and control.

1.5.2. Telephone Control

Telephones are the most ubiquitous form of appliance in the world. Today they outnumber all computer keyboards, mice, printers, and other computer peripherals combined. They come in myriad different forms, from sophisticated office sets, to home sets, to pay phones, to small wireless telephones. As such, they represent an exciting new category of computer peripheral. This aspect of CTI involves the ability, of an appropriately enabled computer, to observe the activity of a telephone set (which buttons are being pressed, what the display says, whether the message indicator is flashing, etc.), and to control the telephone set (updating the display, simulating the press of buttons, turning on lamps, etc.).

1.5.3. Media Access

The third aspect of CTI is media access. This also tends to be the aspect of CTI that is subject to the most confusion. *Media access* in CTI refers to the ability of a computer to get access to the media stream associated with a particular telephone call (that can be specifically observed and controlled through call control). Computer technologies, that manipulate the media stream once it is accessed, are not themselves part of CTI. For example, many computers support fax software that can interpret a fax document and display it on the screen, or transform an electronic document into a form that can be sent using a fax modem. This capability is not a CTI capability. In this example, CTI comes into the picture when the time comes to place (or answer) a call that is to be used to send (or receive) a fax. When the call is established, media access functionality is used to "bind" or associate the call with available fax modem functionality so that the computer's fax software can take control of the fax modem and send (or receive) the fax.[1-2] Likewise, sound streams from the call can be recorded by the computer, sound can be played down the phone line, speech recognition and text-to-speech functionality can be applied, digital data can be exchanged, etc., all given the availability of appropriate media access resources in the telephony products and the corresponding software in the computer.

1.6. Ubiquitous CTI

The field of computer telephony integration has now reached maturity. This has been marked by three key milestones:

- Interoperability – through standard CTI Plug & Play protocols

- Power – though easy-to-use software tools

- Access – through low-cost multimedia communications hardware

The software industry's evolution to better graphical user interfaces, combined with the efforts of early CTI developers, has led to many powerful applications and software tools that can be applied in the construction of CTI solutions. Innovation in multimedia communications hardware products has brought about a new generation of products that make connecting computers to networks and telephone lines easy and inexpensive.

Despite the obvious potential for CTI technology, despite the work by the pioneers of the CTI industry over more than a decade, and despite the tremendous excitement surrounding each new hardware or software technology that promised ubiquitous CTI, the use of CTI technology has only now begun to grow dramatically. Overall, the principal barrier to dramatic growth of the CTI market has been the lack of interoperable products. Without this interoperability, developers must specialize their products for very precise configurations, and customers cannot rely on anything they purchase

1-2. Fax media access — A complete example for fax media access, illustrating how a fax is received from a CTI perspective, is presented in Chapter 6, section 6.11.

and integrate themselves (as is the normal practice for mainstream computer products). There are many, many ways to differentiate CTI products—but interoperability should not be one of them.

Of the three milestones in CTI that have been achieved, interoperability is the most important.

1.6.1. Interoperability

 CTI interoperability is best defined in terms of two operational requirements. Specifically, interoperability has been achieved when CTI hardware and software products from any combination of vendors allow the following:

1. Customers can assemble and upgrade their CTI systems in any fashion using CTI hardware and software products from any combination of vendors. For example, someone can use the same CTI software on the computer at home (connected to an analog residential telephone) and the computer at the office (where there is a digital feature phone desk set). The software takes full advantage of all the capabilities in each location. The customer can even substitute a new CTI server at the office without having to make any changes to the telephone system or application software.

2. Users of mobile computing devices (laptop computers, personal digital assistants, etc.) can take their devices, and all their software, from one work location to another and still take full advantage of the telephony capabilities in each location, regardless of the combinations of products involved. For example, someone can use the CTI software on her laptop with a pay phone at the airport, the hotel room phone, and the phone at a client's site, as well as her cellular phone.

Interoperability standards for telephony have been slow to emerge because in order to satisfy the implications of these two requirements, they had to simultaneously address all the following needs:

- Plug-together, out-of-the-box interoperability

  The true measure of success is that CTI interoperability standards allow customers to simply buy the CTI products that best fit their needs, plug them together, and have them work—the first time.

- Deterministic and robust operation

  People expect flawless operation from their telephone systems, often holding them to a much higher standard than computer systems. No vendor or customer is willing to trade interoperability for reliability. Satisfying this need and the previous one simultaneously is among the most significant challenges. This necessitates defining *normalized behaviors* so that software can be programmed to anticipate well-understood telephone system behavior.

- Platform independence

  The increasing diversity of computer technology—the blurring of the very meaning of the word "platform"—means that complete interoperability specifications must not make assumptions about the type of computing technology being used.

- Support for arbitrary system configurations

  The virtually unlimited range of uses for telephony means that interoperability specifications must be applicable to arbitrary CTI system configurations that reflect both existing usage patterns and those that cannot be foreseen.

- Compatibility with existing products

  Perhaps the biggest challenge of all, interoperability specifications must minimize the efforts required of vendors, developers, and customers to migrate their earlier products to take advantage of the specifications. This means that specifications must anticipate future products and maximize the reusability of existing components.

- Appeal to the entire computer and telephony industry
  The community concerned with CTI technology represents a superset of the entire communications and computer industries. This includes telephone system and telephone device vendors, software application vendors, server and client platform vendors, and those who comment, analyze, and provide consulting to all of these groups.

Fortunately, every major telephony and computer vendor has invested heavily in contributing towards interoperability goals. Each industry group that has tackled these challenges has moved the industry closer to fully satisfying the requirements; each group has built upon and added to previous efforts. These industry efforts have included the following:

- Telephone Manager (Apple and partners)

- CallPath (IBM)

- CSTA (ECMA)

- TAPI (Microsoft and partners)

- TSAPI (AT&T, Novell, and partners)

- CTI Encyclopedia (Versit and partners)

This book provides a complete road map to the state of the art in CTI, based on the latest interoperability specifications to emerge from the industry. It is based on the de facto industry standards for:

- Telephony terminology

- Telephony operational model, features, and services

- Normalized operations

- Protocols

- Configurations

- Programmatic interfaces (APIs[1-3])

1.6.2. Three Phases of CTI Evolution

Consistent with the iterative development of interoperability specifications for telephony, and the maturing of technology in the overall information technology industry, there have been three distinct phases in the evolution of CTI technology:

1. Custom systems

2. APIs for everybody

3. Protocols for systems, APIs for applications

Each phase represents a paradigm shift in the approaches used to develop CTI solutions. Each shift was based on changing economics, technology, and priorities in the telephony, computer hardware, and software industries. The products and practices associated with each phase will continue to exist and interoperate for some time to come. The economics and opportunities associated with developing products for the third phase, however, will mean that more and more industry investment and activity migrates to this approach.

This book deals with all three phases and is intended to help anyone who is considering the alternatives to choose the best approach for meeting their own needs.

First Phase: Custom Systems

The first phase in the evolution of CTI involved custom development. Anyone who wanted to integrate a telephone system and a computer system had to work directly with the telephone system vendor to obtain that vendor's proprietary (and often product-specific) CTI interface. These CTI interfaces ranged from specifications for how a

1-3. API — API literally stands for *Application Programming Interface*. In general usage, it has come to refer to any programmatic interface that allows two independent software components to interact. It is not limited to application software programming. To avoid confusion, this book uses the term *programmatic interface* to refer to generic interfaces between software components. The term *API* is used here to reflect historical usage.

computer could be interfaced to the telephone system in question, to special APIs for specific operating systems. This is illustrated in Figure 1-6.

Figure 1-6. The first phase: custom systems

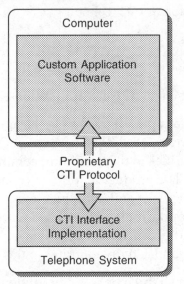

The economics of this arrangement were very poor. In most cases, customers themselves (or systems integrators working on behalf of customers) were developing vast quantities of highly specialized software that was hard-coded to a very specific telephone system. Rarely could this software be reused by others who had the same telephone system. Furthermore, should customers ever need to upgrade their telephone systems, they generally needed to rewrite their CTI software.[1-4]

In instances where telephone vendors had invested in some type of custom API, it invariably was on the wrong operating system for many customers, and was in any case only a slight improvement in the

---

1-4. Conspiracy? — Those who like to believe in conspiracy theories have suggested that the lack of CTI interoperability was a direct result of telephone system vendors seeking to "lock in" their customers. A better explanation is simple supply-and-demand economics.

situation. The custom API still required that the customer develop software specialized to a very specific telephone system. In addition, the telephone system vendor had to keep track of all the popular operating systems. They had to split their development resources between maintaining existing custom API implementations (as new versions of supported operating systems were released), and trying to develop or adapt ("port") the API for new operating systems as they appeared.

Both of these arrangements severely limited the potential size of the CTI industry for a very long time.

Second Phase: APIs for Everybody

The second phase of CTI evolved from the first phase, but it represented a tremendous leap over earlier approaches. It involved APIs that were independent of the telephone system.

The phase-two approach was based on the prevailing philosophy that, given a small number of pervasive operating systems to support, each development platform would provide application developers and telephone system vendors with APIs.[1-5] Each group then could develop the necessary software components to build a complete solution without being dependent upon one another. This is illustrated in Figure 1-7.

The intent of the new approach was to minimize the work necessary for software developers writing CTI applications. Thanks to a single, stable API for a given platform, application software can, in theory, run independently of any specific telephone system or product. This allows software developers who sell shrink-wrap software (not just customers and systems integrators) to develop CTI software. The result should be more applications (so the thinking went), which in turn should mean a more attractive incentive for telephone system vendors to develop the necessary CTI interfaces for their products.

1-5. Telephony APIs — The first mainstream telephony API was the Telephone Manager, developed by Apple in the late1980s. IBM, Microsoft, Novell, Sun, and others were quick to follow over the next few years.

Figure 1-7. The second phase: API layering

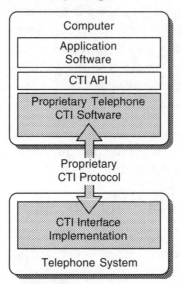

This approach was not perfect, however. While it addressed the needs of software developers, it did little for telephone system vendors relative to the first phase. Telephone system vendors still had to write code for all of the popular operating systems, and they still had to rewrite those pieces of code every time a particular operating system was revised. From their perspective, at a technical level, the only real difference between the first and second phases was the fact that they no longer had to take responsibility for the design of the API.[1-6]

Another challenge associated with the second phase is that solutions built using this approach tend not to be as robust or reliable as desired. The reliability that people take for granted in the world of telephony is in large part the result of international standards that define, in precise terms, the protocols for the ways systems interact. APIs, unlike these protocol definitions, are open to much greater levels of selective interpretation. Applications that use a particular standard API might not work with certain telephone system implementations supporting

---

1-6. API design — Despite the fact that all CTI API developers worked with telephony vendors when designing their APIs, some telephony vendors did not consider this loss of design control to be a positive thing.

that same API. This is in part because the behaviors of the telephone system software are not normalized. As it was for phase one, in these cases the applications must still be specifically modified to work with a particular telephone system. (Generally, however, there is less of this work to do.)

Third Phase: CTI Protocols for Systems - APIs for Applications

While the second phase was characterized by its focus on the needs of application developers, the third phase in the evolution of CTI is characterized by its focus on the needs of telephone system vendors and customers. It involves improving on phase two by addressing the reliability issues and eliminating the key bottleneck, that is, the effort required by telephone system vendors to implement and maintain all the different APIs for all the different operating systems. Both of these objectives are accomplished through the addition of standardized CTI protocols to the APIs developed during phase two.

Figure 1-8. The third phase: CTI protocols

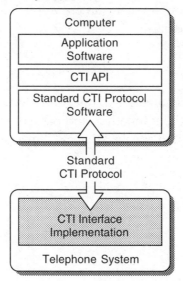

27

In the third phase, telephone system vendors are responsible only for implementing software that runs internally on their own products. The only interaction with application software is through standardized CTI protocols. This is illustrated in Figure 1-8.

The result is that telephone system vendors no longer have to be concerned with how many operating systems they support and which operating systems their customers are using. This also allows their telephony products to be used by devices that don't even have operating systems (or CTI APIs in the conventional sense). Freeing up their resources from platform-specific development projects allows telephone system vendors to focus on providing richer functionality and more robust operation.

The benefit for customers and application developers is that the normalized protocols provide a much higher level of reliability and robustness using the existing operating system-based APIs. Application developers don't have to "special-case" specific telephone systems (unless they want to take advantage of special and unique features,)[1-7] and customers then can use applications without having to worry about compatibility.

1.6.3. CTI Plug & Play

The term *CTI Plug & Play* refers to the ability to take two CTI products out of the box, hook them together, and have them work together without having to install special "driver," "mapper," or other proprietary software in either.

1-7. Vendor specific extensions — Vendor specific extensions are defined in Chapter 4. They refer to unique capabilities supported by a telephone system beyond the standard superset.

CTI Plug & Play[1-8] is made possible by the CTI protocols of the third phase. The use of CTI protocols allows one device to negotiate automatically with other devices to determine their CTI capabilities, so that no effort is required beyond instructing one to connect to the other.

In the paradigm of the second phase, telephone system vendors have to develop special pieces of software[1-9] that allow a computer to work with the CTI interfaces on their products. If these pieces of software are not installed in the appropriate devices, or do not exist, the products simply will not work with one another.

CTI will be truly ubiquitous when all CTI products are CTI Plug & Play.

1.7. CTI Technology, Products, and Solutions

In this book we differentiate among the terms "CTI technology," "CTI products," and "CTI solutions."

The term *CTI technology* refers to the specific technologies that provide the basis for linking computing and telephony technology. The term *CTI products* refers to products that implement CTI technology and make it accessible to CTI solutions. Finally, the term *CTI solutions* refers to the application of CTI technology (by taking advantage of CTI products) in a specific product or working system that combines any number of other communication and collaboration technologies.

1-8. Plug and play — It should be noted that CTI Plug & Play is different from Microsoft's Plug and Play. A device that supports CTI Plug & Play will work with any other CTI Plug & Play device. A device that supports Microsoft Plug and Play will work with a PC that is running the correct version of the Windows operating system and has the correct driver installed.

1-9. Telephony vendor software — The names of these pieces of telephony vendor software differ, depending on the platform. They include *telephone drivers*, *telephony service providers*, *telephone tools*, and *mappers*.

CTI products are always necessary either to support, or be part of, a working CTI solution. Not all CTI solutions are CTI products, however; a CTI solution does not necessarily make CTI technology accessible to other CTI solutions.

Computer-based voice processing products provide an excellent example to illustrate these distinctions:

- Computer-based voice processing systems are CTI solutions because internally they incorporate CTI technology in the form of computer-controlled call processing and media access.

- A voice processing system may take advantage of additional CTI technology provided by another CTI product in order to simplify its own design and/or to get access to a richer telephony feature set.

- A voice processing system also is considered a CTI product if it makes its CTI functionality (either its own technology or that of CTI products it is using) available to other CTI solutions.

1.7.1. CTI Solution Categories

CTI solutions often are grouped into a number of basic solution categories. While these categories generally overlap in arbitrary ways and incorporate other categories, they do make it easier to refer to individual CTI solutions as being used for a particular purpose.

The solution categories referred to in this book are briefly listed below. These explanations are further expanded throughout the book:

- Call accounting
  Call accounting applications involve using software to track information about individual calls (who was called, when the call was placed, the length of the call, etc.) in order to track telephone usage, recover costs, bill for services, reconcile telephone bills, and more. Call accounting applications are

generally easy to justify in environments where telephone use is intensive and time is billed. These products tend to pay for themselves quickly.

- Auto-dialing

  Auto-dialing refers to automating the making of telephone calls. Auto-dialing allows a computer user to simply indicate a desire to talk to someone; the CTI software takes complete care of the process. This can include looking up the right person, finding an appropriate telephone number, determining the best way to call the person (factoring in the current location time zone, etc.), dialing the number and associated billing information, retrying as necessary, and indicating its progress throughout the process. While arguably among the most basic of all CTI solutions, auto-dialing is the most universal in terms of appeal and has the potential for tremendous time savings.[1-10]

- Screen-based telephony

  Screen-based telephony refers to the use of application software that presents a telephone user interface on a computer screen and acts as an alternative to the user interface provided by a physical telephone set. This virtual telephone is usually (but not always) more functional than the telephone set that it supersedes. More important, though, is that it can be customized to meet the specific needs and usage style of a particular individual, who might need more or less functionality than is otherwise available through the telephone set. The functionality of screen-based telephony encompasses auto-dialing and adds support for other

1-10. Auto-dialing time savings — Dialing a telephone number manually (including the time to look up the number if necessary and provide any applicable billing information) can take from 15 seconds to a few minutes. In contrast, the time needed to instruct a CTI application to call a person might be 1–10 seconds. Given the thousands of telephone calls made by computer users each year, the potential time savings are staggering.

telephony features and services. A fully featured *screen-based telephone* application, *SBT* for short, provides access to every telephony function available on a given telephone system.

- Screen pop

    Screen pop solutions involve having a computer system present (or "pop") information pertinent to a particular incoming telephone call onto an individual's computer screen, just as that telephone call arrives at the individual's telephone. This allows the person to prepare for the call even before it has been answered, or even to choose not to answer the call based on the information presented.

- Programmed telephony

    Programmed telephony refers to the broad range of CTI solutions that involve unattended computer interaction with telephone calls. In contrast to screen-based telephony which involves creating an alternative telephony user interface for a human actively managing a call, programmed telephony delegates to a computer the responsibility for interacting with telephone calls. In fact, most applications of programmed telephony can be thought of as creating a user experience for the person at the other end of the phone call. Programmed telephony involves CTI software that allows rules for individual call disposition to be established (or *programmed*). Programmed telephony applications can be as simple as an application that rejects all calls from a certain list of telephone numbers, or as complex as interactive voice response systems that take messages, locate people and information, and redirect calls.

- Voice processing

    Voice processing refers to a subset of programmed telephony solutions that involve using media access to interact with human callers in some fashion. Any application that interacts with callers by playing messages, recording messages, working with speech information, and detecting or generating tones is referred to as a *voice processing solution*.

- Call routing

  Call routing solutions belong to a class of programmed telephony solutions that automate the delivery of calls to selected individuals. Calls can be routed based on associated information provided by the telephone system, or on an actual interaction with a caller using voice processing in some fashion.

- Call screening

  Call screening solutions involve using CTI technology to filter calls and handle them differently, depending upon who is calling or what they are calling about. Call screening solutions may involve screen-based telephony, programmed telephony software, or a combination of the two. *Attended call screening* is just a screen pop solution in which a screen-based telephony software user actively decides to accept or reject a call before the call is actually answered, based on information from the telephone system that indicates who is calling. On the other hand, *unattended call screening* involves routing software that is set up to redirect calls automatically on some basis. For example, an application could act as a private secretary on behalf of a user. It could answer each call, capture information about the purpose of the call, pass only urgent calls on to the user, and take messages in all other cases.

- Auto-attendant

  An *auto-attendant* solution uses voice processing, in lieu of a human operator or attendant, to interact with callers and direct their calls to the desired person.

- Information retrieval

  Information retrieval solutions are voice processing solutions that allow callers to track down and retrieve information without having to interact with a human clerk. Examples include access to prerecorded messages, access to information stored in databases (such as retrieving a bank balance or order status), and retrieval of documents that can be returned via fax or electronic mail.

- Fax-back
  Fax-back is an information retrieval solution specifically involving the retrieval of faxable information. All information requested is provided to the caller in the form of one or more fax transmissions to a selected fax number.

- Personal agent
  A *personal agent* is a piece of programmed telephony software that can act as an autonomous agent for a given computer user. Typically it utilizes voice processing to interact with callers on a user's behalf. The personal agent may provide call screening, if the user is present, or will handle calls independently, if he or she is away. A personal agent might provide any or all of the programmed telephony capabilities described above in the process of handling a call.

- Call center
  When a team of two or more (potentially many hundred) individuals in a particular location handle calls on a dedicated basis, the result is a *call center* (and the individuals are referred to as *call center agents*). A call center may be just inbound (such as a hotel reservation call center) or just outbound (such as a telemarketing organization), or it may be a pool of call center agents who handle both kinds of work.

- Distributed call center
  A *distributed call center* involves all of the functionality and activity of a call center, but the call center agents themselves may be working from two or more locations. A distributed call center might involve a single pool of call center agents working two different time zones, overflowing calls from one to the other at peak times. A distributed call center could involve having each call center agent work from his or her own home, so that no central work location even exists.

1.8. Conclusion

In this chapter we have explored the motivation and context for CTI, identified the three areas encompassed by CTI (call control, telephone control, and media access), and reviewed the evolution of CTI in the context of interoperability. We have also learned to differentiate CTI technology, CTI products, and CTI solutions.

In the next chapter we look at a number of ways in which CTI technology can be applied to build customized CTI solutions in a wide variety of circumstances.

2.
CTI Solutions and Benefits

Opportunities to apply CTI technology and techniques are as infinitely varied and numerous as the number of people and organizations that use telephones. In fact, the primary benefit of CTI technology is that it allows telephone systems to be viewed as solutions that can be tailored precisely to meet the specific needs of market groups, specific customers, and most importantly, to specific individuals.

While the existing diversity of telephony products is staggering, this diversity actually reflects the fact that a huge demand exists for telephony solutions that address an almost infinite variety of needs. It is really an indication of the much larger opportunity that exists when telephony features and services can be more easily brought to bear on the needs of an individual or organization.

CTI makes customization of telephony systems possible because it:

- Opens up access to telephone systems using the well-established flexibility, programmability, and customizability of off-the-shelf computer technology;

- Allows direct integration with an organization's business systems or an individual's work flow tools; and

- Creates many new opportunities for vendors, developers, and individuals with special skills, knowledge of needs, and awareness of individual preferences to add customized functionality to a telephone system.

In short, CTI brings to the world of telephony the kind of revolution that personal computers brought to the world of computing: the ability to get more value out of technology by empowering more people to participate in the development of complete solutions.

2.1. Who Benefits from CTI? (The CTI Value Chain)

The benefit of CTI technology for any individual user or customer of a CTI solution is the functionality achieved when various components and services are integrated to form a solution that is optimal. This ability to obtain different parts of a solution from different sources, and to have someone assemble it in a tightly integrated fashion, is at the heart of both what makes CTI valuable, and what has traditionally made it challenging.

By definition, a complete CTI solution involves the integration of hardware, software, and services from different vendors. This collection of participants in the final CTI system is known as the *CTI value chain*. Each tier in the value chain represents an important part of the final solution; CTI creates new opportunities for the companies and individuals that provide products and services at each layer. This concept is illustrated in Figure 2-1. Table 2-1 (at the end of the chapter) provides a summary of the tiers in the CTI value chain.

2.1.1. Telephone Network Providers

Telephone network providers offer services in the form of access (wired or wireless) to the worldwide telephone network. They also provide optional telephony features to their customers (subscribers) through their portion of the telephone network.

Figure 2-1. CTI value chain

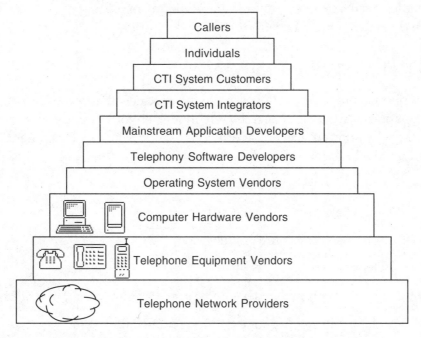

Only a small fraction of the features and services available traditionally are used by the average customer. This is in large part due to the fact that customers have had no good way of accessing much of this functionality, given the simple telephone set's poor user interface for handling many of these capabilities. (Chapter 3 explains the fundamental concepts behind telephone networks, Chapter 4 explains the common features and services, and Chapter 5 describes the forms in which these services are offered by telephone companies.)

CTI represents a big opportunity both for traditional telephone companies and for new telephone network providers that just now are entering the business because of industry deregulation worldwide. In order to sell more telephony features and services, these companies have to make services more useful and easier to use. At the same time, these companies don't have the expertise or might be legally unable to develop the other pieces of a complete solution. However, the concept of the CTI value chain means that they can work with others, who do have the appropriate expertise or products, to build the solutions their customers need. Not every telephone company will pursue CTI solutions actively as a means of differentiating or adding value to their offerings, but all will benefit from it.

2.1.2. Telephone Equipment Vendors

Telephony equipment vendors make the products that connect to the telephone networks managed by telephone network providers, as well as the telephone equipment of other vendors. Telephone equipment vendors include companies that develop, manufacture, and sell:

- Consumer telephone sets and systems

- Business telephone sets and systems

- Cellular and other wireless telephone sets

- Pay telephones

- Computer peripherals designed for interfacing to telephone lines

- Special-purpose telephone and telephony equipment

Telephone equipment vendors have a great deal to gain from CTI. In the past, they have had to compete with one another by staying abreast of the latest advances in telephony technology, differentiating through industrial design, and trying to build telephone systems with the broadest possible appeal. Like the telephone companies, however, their challenge has always been to find ways to let more customers actually take advantage of the features they have built into their products. In most environments, few individuals ever use the full functionality of their telephone system because it isn't tailored to their needs and it isn't easy to use. (Chapter 4 explains the features and services implemented by telephone system vendors, Chapter 5 describes the forms in which these are implemented in tangible telephony equipment, and Chapter 6 reveals how these vendors open their products to expose CTI interfaces.)

By supporting CTI, telephone equipment vendors are better able to meet the individual needs of customers by taking advantage of the other players in the CTI value chain. CTI allows much easier access to a telephone system, so more potential customers actually will be looking for one that offers a rich set of telephony features and services because they will be assembling a CTI solution to take advantage of them. CTI effectively unleashes the latent power of telephone systems, and the result is a bigger and healthier market for telephony equipment in all categories.

2.1.3. Computer Hardware Vendors

Computer vendors play a critical role in the CTI value chain because they build the platforms on which all of the software components in a CTI system run. Computer vendors include companies that develop, manufacture, and sell personal computers, server machines, and multi-user computers, as well as consumer electronics information appliance products and PDAs. They are able to add value to their products through CTI by:

- Supporting slots, ports, and connectors that allow for connection to standard LANs and CTI peripherals;

- Including an operating system with CTI support;

- Building in CTI functionality; and

- Bundling CTI applications.

All of these efforts to bring value to a computing platform through CTI rely on all the other parts of the CTI value chain. For a customer actually to be able to take advantage of a given computer product in a CTI solution, the products and services represented by the other tiers are essential. The computer vendor is relying on telephone network providers and telephone equipment vendors to provide services and products that can be accessed through standard CTI protocols, or with appropriate software. They also rely on the vendors above them in the chain to create the CTI software components that will appeal to customers. (Chapter 7 illustrates how CTI systems are configured by assembling telephony and computer hardware components.)

2.1.4. Operating System Vendors

Operating system (OS) vendors develop the system software that runs most mainstream computer systems. Operating system software provides access to the facilities available on a given computer system, allows applications to be launched, and provides a means for applications to make use of system-wide capabilities.

In most cases, the operating system and not the computing hardware represents the environment targeted by the application software developer. As described in Chapter 1, most operating system vendors have developed programming (or programmatic) interfaces that allow application software running on a given platform to take advantage of appropriately configured and connected telephony functionality. (These interfaces are further discussed in Chapter 8.)

Operating system vendors are much like computer hardware vendors in that they must rely on telephone network providers and telephone equipment vendors to support their proprietary APIs, or to support standard CTI protocols that can be used with these APIs. They also rely on application developers to use their telephony architectures, tools, and interfaces to build products that make a given operating system useful to customers building a CTI solution.

2.1.5. Telephony Software Developers

Telephony software developers are a special category of developers that produce telephony-specific applications. These are applications that take direct advantage of telephony features and services; their primary reason for existence is to extend the functionality of a telephone system or product in some way. The applications produced by these developers are highly specialized towards telephony, however because these applications act as the primary telephony application installed on a given computer, even the most casual user of a CTI system will require one. (Chapter 8 describes the software configuration of CTI systems.)

An example of a typical CTI application is a program that presents a user interface providing all of the functionality (and more) available through the buttons, lamps, and other components on a telephone set.[2-1] Another example of CTI software is a so-called "firewall" server program that allows multiple computers to safely share a single CTI interface on a piece of telephony equipment.

Telephony software developers exist only because of CTI technology. As CTI becomes ubiquitous, the task of writing CTI software will be easier and the market will be larger. CTI developers are able to take advantage of the layers provided by all of the others in the CTI value chain. Like everyone else in the value chain, they are specialists in their area. They are able to thrive only because of the opportunities that result from the access made possible through CTI interfaces, and because of the modularity that results from the division of responsibilities implicit in the CTI value chain.

2.1.6. Mainstream Application Developers

Mainstream application developers also can play an important role in CTI solutions. These developers are the software vendors responsible for writing the popular applications that justify most computer purchases today. These application programs represent the mission-critical applications that everyone relies on day to day to manage their businesses (or their personal lives). A few examples include:

- Calendaring applications

- Contact managers

- Database applications

2-1. Screen-based telephone application — An application that presents an alternative user interface to the telephone is referred to as a *screen-based telephone* or *SBT*. This is discussed further in Chapter 8.

- Accounting applications

- Electronic forms applications

- Spreadsheet applications

- Time and billing applications

- Office management software

- Messaging software

- World Wide Web browsers

The focus of the work performed by mainstream application developers is not on CTI. These developers specialize in solving other problems, yet in many case their applications can be used with dramatic results as part of a CTI system. In fact, CTI's ultimate value for most computer users is integrating these mainstream applications, which they use for the majority of their daily work with their telephone systems. (Operating systems vary in the level of support provided for applications in this category. See Chapter 8 for a complete discussion of CTI software configurations.)

2.1.7. CTI System Integrators

CTI system integrators are a special breed of developer; they assemble CTI systems on behalf of their clients. This involves assembling and integrating the necessary components from each layer in the value chain, and providing whatever configuration and customization of application software is necessary.

Like telephony software developers, CTI system integrators credit their existence to the maturation of CTI technology. They rely on the modularity of CTI product implementations to be able to apply their talent and expertise to the needs of many clients cost-effectively, while customizing each CTI system they construct to specific individual requirements.

2.1.8. CTI System Customers

Regardless of an organization's size (from a home business to a global enterprise), or its objectives (for-profit or not-for-profit), a CTI system brings a multitude of very significant and measurable benefits. These include (but certainly are not limited to):

- Saving money

- Making more money

- Increasing customer service and customer satisfaction

- Increasing internal efficiency

- Projecting a more professional image

- Increasing employee morale

Customers acquire a CTI solution either by having a CTI system integrator do the work, or by assembling and integrating the CTI components themselves. Even after a CTI system integrator has completed an installation, the customer will be able to continue to build on and extend the CTI system as needs change.

The flexibility offered by a CTI-based approach to telephone system design is one of the things that makes it very attractive to customers. Modular CTI technology allows new components, new software, new services, etc., to be added to a solution without having to replace the entire system. Another key benefit of a CTI solution is the opportunity it affords to make changes very quickly in reaction to fast-changing business needs and requirements.

The scenarios in the remainder of this chapter further illustrate the range of benefits that customers can realize by applying CTI technology.

2.1.9. Individuals

Individuals are among the biggest beneficiaries of CTI technology because it potentially allows customization at the individual level. Beyond the solutions that may be developed to optimize work flow from an organizational perspective, modular CTI technology allows an individual to further customize the working environment. By obtaining complementary pieces of software needed to support a particular work style and activities, and by customizing the interface and operation of that software, a given individual can shape the CTI solution to his or her needs and preferences.

This ability to address simultaneously the needs of the organization and the individual through CTI technology is an exciting and worthwhile dimension to these solutions. The scenarios in this chapter illustrate this dual benefit.

2.1.10. Callers (Customers, Colleagues, and Friends)

A fact often overlooked is that a CTI system serves not only its owner, but—just as important—also serves those who interact with the owner through the telephone. Whether the CTI system in question is in a home, small business, or enterprise, callers will benefit from the greater responsiveness, efficiency, and professionalism made possible through a CTI solution.

For example, with CTI technology customers are much less likely to be told, "I'll try to transfer you now, but if I don't get this right please call back." Individuals empowered by the enhanced functionality of a CTI solution will present to their callers a better, more professional interaction. Callers also can benefit significantly from CTI solutions such as automated attendants, interactive response software, and telephony agents that help them get what they need in the absence of a person with whom to speak.

2.2. CTI Solutions

This chapter introduces a number of CTI solutions that are typical of the ways that the technology can be put to work in a wide variety of environments. The scenarios presented, and the use of CTI technology that is described, are intended as examples to illustrate how certain CTI features, functions, and services may be applied, and to stimulate ideas for how you may wish to apply CTI technology in your own environment. These scenarios will provide context as you read this book and become familiar with CTI technology and the underlying telephony functionality that can be exploited with its help.

The CTI solution scenarios described in this chapter are:

1. Screen-Based Telephony

2. Mobile CTI

3. Power Dialing

4. Personal Telephone System

5. Personal Telephone Agent

6. Interactive Voice Response (IVR)

7. Help Desk

8. Call Center

Each scenario is presented in terms of the challenges faced by a particular hypothetical individual or organization before applying CTI technology (if applicable), and how the implementation of a CTI solution has transformed day-to-day activities. In Chapter 9, armed with knowledge of the technologies and functionality provided by each layer of the CTI value chain, you will see the actual implementation of the CTI system for each scenario.

2.3. Screen-Based Telephony

Andrew, an account executive for a large public relations agency, realized that he and his assistant were not getting the full benefit of their company's telephone system. Despite the personal computers and advanced digital phones on every desk, they were still doing business the same old way. Andrew was an intensive telephone user and spent most of the day talking to editors and his clients. The fact that the telephone system wasn't easy to use was getting in the way. The final straw came when he overheard his assistant saying, "I'll try to transfer you now, but in case this doesn't work, please call back and ask for extension 40220." Everything changed when he put a CTI solution to work.

Andrew connected both his and his assistant's phones to their personal computers and installed screen-based telephone application software. This alone was a dramatic improvement. He then could start taking advantage of the advanced features of the telephone system that heretofore never were easy enough to actually use. Transferring and conferencing calls became a snap and now always work; callers are never cut off or left waiting to be connected. In addition, features of the telephone system of which he'd been unaware, such as call parking, are now useful tools in his day-to-day work.

Voice mail is now easier to access as well. His screen-based telephone software alerts him when there are messages waiting; all he has to do to retrieve them is click on the application's voice mail button. The application dials the voice mail system and logs in automatically. By clicking on buttons on his computer screen, he can listen to each of his messages, save and delete messages, rewind and fast-forward individual messages, and even change his greeting—something he had never figured out how to do by himself.

Another benefit Andrew noticed was reduced training time for new employees. When his assistant went on vacation, her replacement was able to use the software after running through a quick tutorial on her first morning.

However, in many ways the real productivity gains have come from hooking the company directory database and his personal contact manager application (an off-the-shelf software product) to the screen-based telephone software. The result greatly streamlines the placing of calls.

To place calls before installing the CTI solution, he had to look up the number of the person he was calling using his contact manager, and then manually enter the phone number on his phone. Looking up the telephone number was the easy part: he had a function key macro to bring up his contact manager software, and he could find entries just by typing the first few letters of the last name. The annoying part of the process was having to dial the number itself. Depending on what area code he was calling, he had to dial 9 or 8 before the number. (He could never remember the rules, so he had a cheat sheet beside the phone.) If his contact was out of the office, he would hang up and repeat the process using the person's cellular phone number, home number, or ultimately the pager number or the person's assistant.

After setting up his CTI solution, all Andrew has to do now to place a telephone call after looking up the person's name is click on the appropriate telephone icon beside the name or press the Enter key on his keyboard. The screen-based telephone software does the rest. It even puts his telephone set into speaker-phone mode so that he doesn't even have to move his hands away from the keyboard. With his new software in place, Andrew can call any contact using an average of five keystrokes. From the time he decides to call someone to the time he hears the phone ringing is less than five seconds, and his hands don't even have to leave the keyboard (or mouse).

In fact, Andrew can dial any number on his computer with just a single gesture of the mouse. Any telephone number that appears in a piece of electronic mail, a word processing document, a spreadsheet, or an Internet page can be dialed just by selecting it and dragging it to the screen-based telephone application. The application figures out the type of number (internal, local, or long-distance) and dials the number appropriately. Andrew is now looking forward to installing similar software on his personal digital assistant so he can perform many of the same functions when he is away from his desk.

In summary, Andrew's CTI solution has allowed him to be much more effective in his job. His clients and press contacts are treated with a high level of responsiveness and sense of professionalism, and from Andrew's perspective the system is more efficient and reliable. He now is making far better use of the agency's substantial original investment in a telephone system and personal computers, despite the small incremental cost of the solution and the short time it took to get it working. Perhaps most important, work is a lot more fun for Andrew and his assistant.

2.4. Mobile CTI

Betty is a sales representative for a sportswear company that specializes in short-run custom manufacturing. She spends all of her time traveling and meeting with prospective customers. Her office literally is her notebook computer and the nearest phone. She does all of her work from hotel rooms, airport lounges, and taxicabs. When she's lucky enough to be at home, she works from her spare bedroom. Her telephone system is whatever is installed in a given hotel room, the credit-card phone in the airport lounge, a cellular phone when she's in motion on the ground, and the phone at her airplane seat when she's in the air.

While Betty enjoyed the pace of her job, loved the travel, and relished the challenge of working with new people every day, this array of different telephones and ways of dialing numbers was the one sore point. In fact, she often delayed returning phone calls for this reason, though the telephone was a very important part of her job. The sales process involved meeting with prospective clients, leaving samples, working out the purchase details over the telephone, faxing or e-mailing a proposal, and getting a confirmation of the sale. When her prospects wanted to reach her with questions, requests for additional information, or were ready to place an order, they left her electronic mail or sent a page. She would call back from wherever she was, answer their questions, take down their requirements, and then follow up with information or a proposal using the customer's preference of fax or e-mail.

Every time she placed a call, sent a fax, or connected to her electronic mail service she had to enter up to 36 digits. For example, to return a call to a customer in San Jose from her hotel in Boston she had to dial the following:

- 9 to get an outside line

- 1-800-MYT-ELCO to call her long-distance telephone company

- 408-555-1234 to call her customer's number

- 666-777-8888-9999 to enter her billing number

If she made a mistake at any point in the sequence, she had to start over. Though she was very good at punching in all of these numbers, the minute or so that it took still distracted her from thinking about what she wanted to say to her customer.

If this weren't bad enough, the numbers to dial also differed depending on where she was and what kind of phone system she was using. This meant that she always had to think about (or actually look up) the right way to dial a call from her current location. For example, in Toronto she had to remember that to call Hamilton (area code 905) to the west was a long-distance call, but dialing east to Pickering (also area code 905) was a local call.

Needless to say, Betty was very excited when her company had a system integrator develop a sales-force automation solution that included a CTI component. The new system dramatically simplified every aspect of using communications technology in her job, and let her really focus on the time she spent with her customers.

With the new system, a typical sequence of events when Betty arrives in her hotel room might be as follows:

- She attaches the modem in her notebook computer to the hotel phone line and tells the software what city she is in and what hotel she is calling from (if it is one where she has stayed before). If it is a new hotel, she enters the information for placing local and long distance calls and saves it for the next time.

- She instructs her computer to retrieve her outstanding electronic mail while she unpacks. The dialing software knows that her Internet service provider has a local number, so it dials that one rather than placing a long-distance call.

- If a new piece of electronic mail from a customer requests a phone call, she simply drags the phone number from the electronic mail to the screen-based telephone application she is using. Placing the call is fast, effortless, and error-free regardless of where she is. She lets her computer dial the number and then lifts the handset to wait for the customer to answer.

- Whenever she calls a customer, a special sales-force automation application developed by the system integrator pops up on the screen. It is aware of whom she is calling because it monitors use of the screen-based telephone application and provides context-based assistance. Before the call is even completed, this software has retrieved all of the information about the customer she is calling, based on the phone number called.

- If during the course of her telephone conversation a customer asks for information about a particular product (such as the price), Betty just presses the "product info" button in the sales-force automation application window and enters the product code. Because the application already knows which customer is on the phone, it can produce the correct pricing information for that particular customer. If the customer wants a hard copy of the information, she selects the "send fax" option; her computer generates an appropriate cover letter to go with the product information and places the resulting fax in queue, from which it will be sent as soon as the phone line is free. The system knows the fax number already because it is in the customer database.

- When a customer wants to place an order, Betty presses the "place order" button. An electronic form is presented with most of the fields already completed, based on the customer context. She completes the rest of the form according to the customer's request. The form then is automatically sent to the customer by electronic mail or fax for confirmation.

- When the customer wants to schedule a meeting, Betty presses the "set up meeting" button and is presented with her calendar. As with the other automated functions, once she agrees on a meeting time, a confirmation letter is automatically generated and sent via fax or e-mail.

The software works in any place she can connect her notebook computer to the phone line. This includes her hotel rooms, the frequent traveler lounges at the airport, her cellular phone, and the phone in the airplane. Betty has already customized the solution to use a commercially available screen-based telephone application that has a smaller window. (The one provided by the system integrator supported many features for use with the company's office system, but because she never visits the office, these features were never active.) Betty is now looking forward to being able to use the infrared port on her notebook computer to control pay phones, and to use simultaneous voice and data (SVD) technology to be even more responsive to her customers.

In summary, Betty's new CTI solution transformed her job. It took the technological drudgery and communications nightmare out of her job and actually put her notebook computer to useful work. Dialing not only is fast and easy, but calls are always placed correctly the first time and always use the least expensive carrier. Overall, Betty enjoys every aspect of her job and her customers see much greater professionalism, consistency, and responsiveness.

2.5. Power Dialing

Chuck works in Accounts Receivable for a division of a Fortune 500 company. Until recently, his week had a very specific, predictable schedule. Four days a week he processed payments and updated customer accounting information in the company's AR software database. Every Thursday evening he would run the past-due accounts report, and on Friday he would spend the day pursuing customers whose accounts were past due. Tracking down outstanding payments was easy once a customer was on the phone: it was usually just a clerical error or oversight that needed to be addressed. Actually reaching the customer was what took most of the time, and leaving voice mail never got results. The routine went as follows:

- Starting at the top of the report, look up the number for the customer and dial it.

- If the number is busy or if it goes to voice mail, hang up and remember to try this customer a little later.

- If the call goes through to a live customer, pull up the appropriate customer information on the computer terminal and resolve the issue.

- Repeat for the next customer in the report.

- When the bottom of the report is reached, start at the top and try all of the people who weren't reached on the preceding pass.

Fridays were very long and tedious days for Chuck, and he didn't look forward to them. All this changed, however, when a friend of his in the IS department connected the accounting computer to a new CTI server and wrote a simple program (based on the logic above) that automated the whole process.

Using the new solution, the computer does all of the dialing for Chuck automatically. In fact, it uses a telephony capability known as *predictive dialing*, which allows the telephone system to place the call automatically on Chuck's behalf and to hang up automatically if it detects a voice mail system or a busy line. Chuck is only involved in the calls that successfully reach a person. When a customer is reached, the appropriate customer record is presented on Chuck's terminal and Chuck's speaker phone is activated.

The new CTI solution makes the process of collecting overdue accounts simple for Chuck. In fact, because the new software does virtually all of the work, it frees Chuck to work on his other AR responsibilities between calls. This means that collections activity can run all week long. (Extra logic was added to make sure that customers were not called more than once per week.) Chuck activates the dialing software while he is doing other work and deals with customers as the system successfully finds them.

Chuck is now looking forward to the new accounting system, which is being developed to run on a network of personal computers. With this system his dialing software will become even more powerful.

In summary, this simple CTI solution has both improved Chuck's job and speeded the collections process, saving the company money. It was relatively simple for someone in the IS department to implement, and it took little time for Chuck to learn how to use it.

2.6. Personal Telephone System

The Morgans are a typical family with a not-so-typical phone system and home computer—but it wasn't always that way. The Morgan home computer is their all-in-one communications appliance. The kids use the family computer for their school work and they particularly enjoy surfing the Web. Dad uses it to manage the household finances and to connect to his company's network to retrieve his electronic mail. During the day, though, it's down to business: Debbie Morgan has a home business and everything revolves around the computer.

When Debbie first started her home business, it wasn't long before she discovered that she couldn't take telephony for granted. The telephone quickly became central to her business. She started out using her home telephone line, but it soon became awkward to keep track of which calls on the monthly bill were business calls and which were not.

Privacy became an issue when business calls came during the evening. She would take the calls in her office, but occasionally the kids would pick up the phone in the kitchen and disrupt her call. Furthermore, with only a single line, she was missing calls from important clients when she was using the telephone line for sending faxes.

That's when Debbie invested a little effort in setting up a CTI solution. In assessing her needs, she concluded that while the home definitely needed a second phone line so she could use one for faxes and another for voice calls, the cost of a third line could not be justified because there was virtually never a time when all three lines would be busy. She opted for a personal PBX (a personal telephone system the size of a computer hard drive) and a second phone line. She rewired the existing phones in the house so that both telephone lines came directly from the telephone company into the personal telephone system. She then connected her fax modem, her office telephone, and the lines leading back to the kitchen and bedroom extension phones into the new telephone system. With this system she got the benefits of powerful CTI features, along with the ability to share two lines among all of the users in the household.

Debbie's customers are located all over the country, but with her screen-based telephone application, they're all just a mouse click away. When Debbie wants to follow up with a client, she just retrieves the right page in her client database and clicks on a dial button. Much more exciting than automatic dialing, however, is the support her new system has for presenting information about a caller when the phone rings—even before she answers. When a client calls, the appropriate page of her client database is displayed and she can decide whether she wants to talk to the person or have her computer take a message. If she answers the call, all the information she needs is already on the screen. In fact, the computer actually acts as a speaker phone, so her hands stay free during a call and she doesn't have to invest in another, separate piece of equipment.

The time and duration of every call is tracked, so it's a snap to sort out which calls are personal and which are business. Client billing is automatic because, with a little bit of scripting, Debbie was able to tie the call-logging application into her time and billing software!

The Morgan computer even is the household's full-time answering machine and fax machine. It sorts out which calls are faxes and which are voice calls and presents both in an on-screen "mailbox." Callers on Debbie's business line are presented with an appropriate greeting; callers on the home line can choose the family member for whom they wish to leave a message. Faxes sent to either line are detected automatically, so that no telephones ring and the fax goes directly to the computer.

To summarize, Debbie Morgan's home business in many ways is made possible by the combination of her personal computer and her personal telephone system. The combination delivers an easy-to-use set of capabilities that are seamlessly integrated, allowing Debbie to focus on working with her clients. Debbie saved a lot of money by using this type of CTI solution rather than buying extra phone lines, a dedicated fax machine, a speaker phone, and separate answering machines. The system also saves Debbie a great deal of time in many areas. She is always working to a tight deadline, so knowing who is calling before answering the phone avoids having to spend valuable time with a nonpaying client. In addition, the logging feature means saving the time that otherwise would be spent sorting through phone bills, and ensuring that every call is billed back to an appropriate client if possible.

2.7. Personal Telephone Agent

Edmund is a very busy person. He is a free-lance photographer who is always on the move. While he is at work his time is split between his office, his studio, and his darkroom. The rest of his time is spent traveling to outdoor locations. Edmund doesn't have a secretary or other administrative support; his personal computer does everything from tracking his schedule to automating his accounting and billing.

The one thing his personal computer wasn't doing for him was handling phone calls. If he wasn't in his office (which was the majority of the time), an answering machine answered the calls. From time to time he would check his messages either by returning to his office or calling into the answering machine from his cellular phone. After a few occasions when he missed out on some business because he didn't get back to an important client quickly enough, he decided to invest in a CTI solution.

Edmund spent some of his spare time setting up a personal agent. Once his system was fully customized, it was able to autonomously handle all of his calls. If he is away from the office or doesn't want to be disturbed, the personal agent answers all calls automatically. The personal agent interacts with each caller to find out who they are and the reason for their call. If a call is indicated as being urgent, the personal agent asks the caller to wait while it searches for Edmund. The software then tracks down Edmund as follows:

- An intercom system connects the office to the studio and darkroom. The personal agent first tries to find Edmund by announcing the caller's name and the subject of the call on the intercom speakers.

- If Edmund doesn't respond, it then tries calling him on his cellular phone, and then at his home number.

- If all of these attempts fail, it takes a message from the caller and sends an urgent pager message to Edmund, indicating the nature of the message that has been taken.

Not only does Edmund have the ability to call into his personal agent to take urgent calls that his agent has tracked him down for, he also can call in to retrieve important information from his personal computer. For example, if he forgets the time and location of his next meeting, his personal agent can read this information to him over the phone, and if he forgets an important document, his personal agent can fax it to him.

In summary, Edmund's personal telephone agent is an example of a very powerful CTI solution that allows anyone to turn his or her personal computer into a full-fledged assistant. In Edmund's case, it gave his photography business a significant edge over his competitors. His business now projects a more professional image, and he is accessible 24 hours a day without having to change his lifestyle or the way he runs his business. This CTI solution saved Edmund money. Instead of hiring a receptionist or answering service, he put his computer to work. An extra benefit is that he now has 24-hour remote access to his computer from any telephone, anywhere, at any time.

2.8. Interactive Voice Response (IVR)

Frances is the vice-principal of a primary school. At the last PTA meeting, a number of parents presented an interesting idea. They wanted to be able to call the school to find out about each evening's homework, special events, snow closures, etc. There were only two problems with this request. The first was that the school secretary already was very busy. The second was that the secretary didn't work past 5:00 PM and the parents needed to be able to call in the evenings.

Frances decided to put the school's administrative personal computer to work at nights. With the help of one of the school's computer whiz kids, she set up the computer to answer the school's phone number at night. At some point each day, each teacher drops by Frances's office and records a message for the parents of the kids in his or her class. These messages include descriptions of the evening's homework and hints the parents can use to help their kids. Frances records any special announcements and then leaves the computer running when she goes home for the day. Parents can call the school at any time, the computer answers, and they can punch in their child's grade on their touchtone phone and hear the appropriate prerecorded announcement from the child's teacher. They can also listen to prerecorded school-wide bulletins, special notices, award announcements, etc. In the event of an emergency, such as a snow closure, Frances is able to dial into the system and leave a special message for parents.

The new system is such a success that Frances already is planning expansion. She wants to add two more telephone lines and make the information accessible over the Internet for those families who have Internet access from home.

This scenario demonstrates that CTI is not just for business, but for any organization that wants to take full advantage of technology to provide a better service to their community.

2.9. Help Desk

Gunther works for the information systems group at a university. The information systems group is responsible for all of the computers and networks in the university and its dormitories. Gunther is a member of the help desk team that is responsible for providing technical assistance when people have problems. Because everyone in the university community uses a computer and these tools are so essential to their academic pursuits, a priority has been placed on making sure that all members of the academic community have all the necessary support to keep their systems running. The university also is extensively networked. Every office, lab, and dorm room has an Ethernet connection to the university's computer network and a telephone connected to the university's telephone network.

The help desk is a telephone number that is answered 24 hours a day by a team of highly trained computer technicians, like Gunther, who can help users troubleshoot their problems.

When the help desk was first established, it was almost immediately overwhelmed with calls. Gunther and the other help desk team members identified the following problems:

- Despite the fact that each technician had a different area of specialization, calls would be delivered to the different technicians at random. Often it was not until a minute or more into the call that the technician was able to determine that the problem might be better handled by someone else. At that point the call could be transferred (and the background information repeated), or the technician could take a stab at the caller's problem.

- After a technician had explained how to a fix a problem, callers preferred to stay on the line while they implemented the solution; they didn't want to risk having to explain the problem (and the fix that was recommended) to a different person if the fix didn't work. The result was that calls that should have taken only two or five minutes actually were tying up a technician and a phone line for as much as half an hour.

- Another reason callers wanted to keep their technician on the line was that waiting times were so long that they didn't want to have to call back again if their problem wasn't solved.

The team decided to let Gunther implement a CTI solution to eliminate these problems. Here's what he did:

- Gunther began by building a simple database, using an off-the-shelf database product, that would track all the information relevant to every problem report. Every technician's computer has access to this multi-user database, and the appropriate record is presented whenever one of them takes a new call. All the information is available to them even if they didn't handle the call that generated the record.

- Gunther set up a voice processing system that greeted each caller to the help desk and asked basic questions about the problem. It could associate the call with an existing request from the caller, or could create a new record in the database.

- Gunther set up some call-routing software that uses the information about each call in the database to queue and then deliver it to the technician who last handled the call if possible (in the case of a repeat call), or to the technician with the correct expertise based on the nature of the problem (if it's a new call or the original technician is not on duty).

- Gunther also set up a simple Web site[2-2] on the university's intranet that provides both answers to the most frequently asked questions (to eliminate the need to even talk to a technician if possible), and also an escalation mechanism if the needed advice cannot be found. The escalation mechanism is an alternative to the voice processing system for inserting a request into the call queue. If a particular computer user's problem is such that the web site can still be accessed, he or she can create a new problem report or indicate an existing one, and request a call to his or her location. When the user's turn comes up, the system automatically calls and connects the call to a technician. In this way, calls are queued on a first-come, first-served basis, but they don't have to tie up phone lines. Web site users also can monitor their positions in the queue so they know roughly how long they'll have to wait.

In the future, Gunther would like to add support for screen sharing so that a technician can use the network connection to look at a user's screen, speeding the diagnosis and the fix.

2-2. Web site — A Web site is a collection of Web pages on an Internet World Wide Web server, or on an equivalent intranet server. A Web page is an HTML document accessible to Web browser applications. These applications translate the document and present it on the computer screen in the form of a hypertext display. In this scenario, the HTML document viewed by each client is created dynamically and is specific to a particular request.

By implementing this system, Gunther was able to make the help desk a much more effective resource in the university community. It saved the university from having to hire many more technicians and provided much more efficient service to callers.

2.10. Call Center

In the case of Henrietta's virtual travel agency, there is no life before CTI. Her business is one that is possible only because of CTI technology.

Henrietta manages a small travel agency with six agents. Competing with the big chains is tough, but Henrietta has CTI technology on her side. It gives her agency the capabilities of a big business with the personal touch of a small business, and the overhead of cyber-business.

Henrietta's CTI solution delivers unprecedented customer satisfaction, streamlines work, and creates a more pleasant work environment for her employees because they all work from their own homes as a distributed call center. The hub of her business is a server and networking equipment in her basement.

Her server hosts an auto-attendant and a customer database, which her agents use to track clients and activity. Each of her travel agents has a computer with a dial-up connection to her server and a telephone for business use. The different agents work at different times of the day, depending upon when demand is highest. When customers call (either to the local number or Henrietta's 800 number), their calls are automatically forwarded to the agent with whom they last talked (if he or she is working). All the information about a given customer is presented on the computer screen before the agent even picks up the phone. If the appropriate agent is unavailable, the caller can be redirected to another agent or can leave a message. The agents all have pagers, so if they are not working when a call comes in they are notified as to how the call was handled.

CTI Solutions and Benefits

Calls that come in overnight (typically emergency calls from stranded customers) are handled by a designated on-call agent. These events are rare, so the agent doesn't have to sit by the phone; he or she merely stays home that night. When such a call arrives, the server will send the call first to the designated agent and, if this does not produce a response, a message is taken and the agent is paged repeatedly until he or she does respond. The emergency service is rarely used, but it is a significant differentiator for Henrietta and allows her to handle corporate clients who would otherwise go to big chains that can afford to staff a call center 24 hours a day.

Another benefit of the CTI system is that it saves money by streamlining outbound calls. Agents just click on the phone numbers for clients, hotels, airlines, etc., and the call is placed instantly. Misdialed numbers are a thing of the past, and calls are always dialed the most cost-effective way available. When Henrietta's agents reach an airline or other travel service provider that has its own multi-step voice processing system (phone maze), their screen-based telephone software is preprogrammed to navigate to the desired service with the click of a mouse.

In the future Henrietta will be growing her business by adding a Web site to her server. By using an Internet Telephony Gateway she'll be able to reach a new market, cyber-customers, and save charges to her 800 number by feeding calls from the Internet into her network.

Henrietta's business is indistinguishable from the advanced distributed call center she has built. Not only does CTI technology make her employees very happy because they can work at home, and improve on customer service and satisfaction because of the personalized treatment they get, it also significantly reduces the overhead of the business and keeps the variable costs to an absolute minimum. Callers are automatically directed to the agent who can help them the most effectively, so calls are shorter and agents can handle more customers.

2.11. Conclusions

In this chapter we have seen the tremendous opportunities and benefits that exist for those who invest in CTI technology. This applies both to customers and to those in the CTI value chain who provide the individual components that make up working CTI systems.

This chapter has also demonstrated the importance of interoperability and modularity in the design of CTI components. The people described in each scenario were able to build and use their CTI solutions because they were able to go to different sources for the pieces that make up the final solution. No single vendor could have provided the complete solution. Each involved the customer integrating multiple components and customizing the solution to meet specific needs.

These scenarios have become possible as a result of the maturation of CTI technology and the associated recognition of the CTI value chain. Given how very compelling these scenarios are, market forces (the economics of interoperability) will encourage vendors to build CTI Plug and Play products so that solutions like the ones described can be realized with ease.

The steps in building or customizing a complete CTI solution from individual components, or for building an individual component that is to be used as part of a CTI solution, include the following:

1. Familiarize yourself with all of the technology, functionality, and value that is added at each layer.

2. Decide which layers you can or want to do yourself, and which ones you'd be better off relying on someone else to provide.

3. Decide what functionality you need to provide in your CTI component, or at each layer in your CTI system.

4. Armed with the resulting knowledge, proceed with the implementation of your system or component.

Regardless of where you fit into the value chain, it's a good idea to be aware of what takes place above and below. Then, and only then, can you be sure that you are taking full advantage of the opportunity that CTI represents to you or your organization. This book provides a thorough insight into the technologies and considerations associated with each layer. As shown in Table 2-1, we will be working our way through the value chain from the bottom up. (This book is designed, however, so that you can go directly to the chapter corresponding to the layer you are most interested in if you wish. References will guide you back to key concepts you might have skipped.)

Table 2-1. CTI value chain

Layer	Role	See Chapter
Telephone Network Providers	Provide telephone network access and services	3,4,5
Telephone Equipment Vendors	Provide equipment for connecting to telephone network	4,5
	Implement CTI interfaces	6
Computer Hardware Vendors	Provide hardware platforms for CTI software	7
Operating System Vendors	Provide software platform for CTI software	8
Telephony Software Vendors	Provide indispensable CTI-specific software	8
Mainstream Application Vendors	Provide mainstream/mission-critical application software	8
CTI System Integrators	Assemble and integrate components for CTI solutions	9
CTI System Customer	Build and/or customize CTI solutions	9
Individuals	Use and customize CTI solutions	9
Callers	Benefit from CTI solutions	

3.
Telephony Concepts

More than a century of innovation and creative product development have created an extraordinarily diverse range of telephony products and technologies. This diversity, like the diversity seen in computing technology, has been very good for customers who have benefited from having a wide array of technology choices. But unlike the computer industry, which always has been reined in by the need for highly functional interoperability and standardization of component interfaces, the telephony industry traditionally has not had to address the need for integrating with off-the-shelf computing technology. (Their traditional worry has been interoperability among telephone networks.) As a result, the diversity, lack of consistency, and rapid innovation have been both a blessing and curse for those challenged with making CTI technology ubiquitous. Organizations such as the ITU, ECMA, ECTF, and Versit, however, have invested heavily in developing interoperability specifications that provide a consistent overall framework, but also reflect the rich diversity of proprietary approaches. A fundamental aspect of these efforts has been in developing a consistent body of nomenclature to replace the arbitrary and proprietary terminology that obscures the many fundamental similarities between implementations.

Telephony Concepts

Despite the tremendous diversity in how telephony equipment vendors actually implement the products and interfaces that allow telephone calls to be placed, the basic concepts are, in fact, universal.

Telephony and Telecommunications

The terms *telephony* and *telecommunications* are frequently confused. Traditionally these words have been used to differentiate between voice telephone networks used for telephony, and other networks (including those based on telephone networks) that are used to move data between computer systems (telecommunications). Over time, the telephone network has evolved into a powerful digital communications infrastructure in which voice is but one type of data. This has led to some confusion because one could model the telephone network as a telecommunications network that supports telephony, or as a telephony network that supports telecommunications. Telephony has more stringent requirements, however, and its functionality is generally a superset of telecommunications functionality; you can always layer telecommunications on top of a telephony infrastructure, but you can't always do the reverse.

 A *telephone system* is a collection of interconnected or related *telephony resources*. Any subsystem, or portion, of a telephone system is itself a telephone system and represents a particular *telephony resource set*. A telephone system may be as simple as an individual telephone or as vast and complex as the worldwide telephone network.

This chapter presents the universal telephony concepts by explaining the various types of telephony resources that can be found in a telephone system and the entities they manipulate. These concepts can be applied to any telephony product by modeling the mechanical, electronic, and software components[3-1] that make up its implementation as a telephony resource set. (Telephony products are discussed and modeled in Chapter 5.)

3.1. Telephony Resources

A complete set of telephony resources can be visualized as shown in Figure 3-1. Each block in this diagram represents an abstraction of a distinct area of telephony functionality. Most telephone systems have only a subset of the telephony resources shown.

Figure 3-1. Telephony resource set

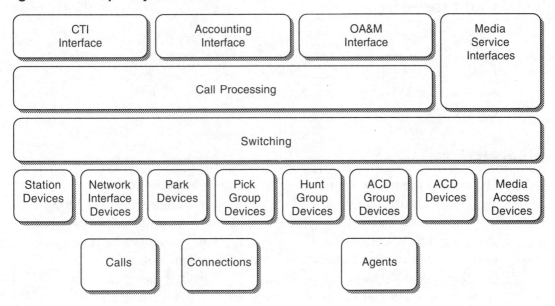

3.1.1. Switching Resources

Switching resources represent the switching function in a given telephone system. Switching refers to the ability to connect telephone calls and is therefore the most fundamental function within a system. Within the switching resources are a finite number of channels that are

3-1. Human system components — To be complete, it should be noted that in addition to mechanical, electrical, and software components, some telephone systems (especially antique ones) include humans to implement one or more of the telephony resources. This is referred to as "putting humans in the system." In most cases humans are considered users of the system and are therefore outside it.

used to convey information from one place to another. The switching resource in a given telephony resource set is the only required resource. In a basic telephone it could be as simple as a mechanical switch that connects the phone to a telephone call. In modeling the worldwide telephone network, this resource represents the ability to connect any two telephones in the world.

3.1.2. Call Processing

Call processing is the "brain" of a telephony system. It manages all the information about how the telephony resources should behave with respect to one another, it accepts commands from the other resources, and it manipulates the resources in order to carry out these commands.

3.1.3. Devices

Devices are the resources in a telephone system between which the switching resource is able to provide connections. They are referred to as *endpoints* and, as we shall see later, devices are essential to the way that we model telephony functions. While there is no upper limit to the number of devices a particular telephone system may have, a functional system has no fewer than two devices that may or may not be connected together by the switching resource at a given instant.

There are eight basic types of devices:[3-2]

- Station devices (telephone sets)

- Network interface devices

- Pick group devices

- Hunt group devices

- Park devices

3-2. Device types — As with all aspects of the abstractions presented in this book, any implementation may invent unique functionality and represent that functionality by extending this model with new types of resources.

- ACD group devices

- ACD devices

- Media access devices

3.1.4. Dynamic Entities

Dynamic entities are abstractions that represent dynamic relationships between resources in a telephone system. Along with devices, these entities are central to the conceptual model for telephony. There are three types of dynamic entities:

- Calls

- Connections

- Agents

Calls, connections, and agents are explained in detail in this chapter.

3.1.5. Interfaces

The last category of telephony resources are *interfaces*, which convey command and status information to and from the telephone system. These interfaces include:

- CTI interface
  The *CTI interface* allows computer technology to interface with call processing in order to send commands and receive status information. This interface is also referred to as an *open application interface* or *OAI* by some vendors. The nature and use of this resource is the principal topic of Chapters 6, 7, 8, and 9.

- Operations, Administration, and Maintenance (OA&M) Interface
  The *OA&M interface* provides a configuration mechanism for instructing call processing and the other resources how to behave when put into operation. It also may provide diagnostic or status information and allow telephony resources to be taken in and out of service.

- Accounting interface

  The *accounting interface* provides a mechanism for tracking the usage of resources in order to appropriately account or bill for their use.

- Media service interfaces

  The *media service interfaces* provide external access (using corresponding media interface devices) to media data such as voice, video, fax, or other data stream types. (These concepts will be explained further in this and subsequent chapters.)

3.2. The Basics: Calls, Devices, and Connections

At the heart of a telephone system's functionality is the ability to make and manage telephone calls, so our exploration of telephony concepts logically begins with the concept of *calls*. *Devices* and *connections* round out the three most important concepts in the world of telephony.

As we shall see, no matter how complex the telephone system, its basic operation can be described in terms of these three concepts.

3.2.1. Media Streams

The simplest way of thinking about a telephone call is to visualize it as a pipe that delivers streams of voice information through a telephone network, to and from a device such as a telephone (Figure 3-2).

Figure 3-2. Voice on a simple telephone call

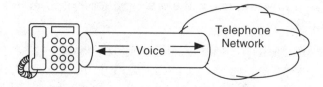

These voice streams are referred to as *media*. In general, this media information is an isochronous stream of data that may be voice or some other type of arbitrary data.

Isochronous Streams

One unique characteristic of the telephone network is that it is capable of delivering data in an *isochronous* fashion. Most data networks are not capable of supporting isochronous data streams.

Isochronous data streams deliver bits of information at a guaranteed, constant rate. For example, your telephone network connection might deliver 64000 bits of information to your telephone every second. These bits are converted to sound, which you hear as a continuous stream of voice information.

In contrast, asynchronous streams deliver bits of information on as-able basis. Data is received in random bunches based on the availability of data and transmission resources. When voice data is transmitted on an asynchronous connection, you might hear pops, clicks, and periods of silence.

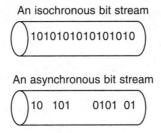

An isochronous bit stream

1010101010101010

An asynchronous bit stream

10 101   0101 01

3.2.2. Control Information

In addition to the media information associated with a telephone call, *control information* must be associated with a call. Sometimes referred to as *signaling*, this information is used by the telephone system to set up, manage, and clear the telephone call. This is illustrated in Figure 3-3.

Control information is just as important as the media stream itself. Without that associated information the system wouldn't, for example, know how to connect the call.

Figure 3-3. Control information on a telephone call

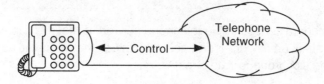

3.2.3. Calls

The abstraction of telephone calls (which are referred to as simply "calls" from here on) consists of a media stream and associated signaling information, as shown in Figure 3-4. Keep in mind that this is just an abstraction; in Chapter 5 we'll see how this abstraction relates to different implementations.

Figure 3-4. Telephone call abstraction

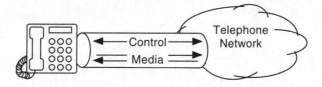

 Call refers to both a media stream that is established between endpoints in a telephone network and all associated control information.

In building CTI systems we are concerned both with call control, which involves getting access to the control information, and with media access, which involves getting access to the media stream.

3.2.4. Devices

Another important element in the abstraction of a call are the endpoints of the media stream.

An endpoint to which a telephone network is able to connect calls is a *device*. In the preceding illustrations, a simple telephone is shown as the device at one end of a call. (The abstraction of devices will be expanded later in this chapter as we look at specific types of devices and further explore the abstraction.)

A functional telephone call involves two or more devices as endpoints to the media stream associated with the telephone call. At certain points in the life of a call, however, such as when the call is being originated or cleared, it may have only one device or no devices at all.

3.2.5. Connections

The relationship between a particular device and a particular call is referred to as a *connection*. If two devices are involved in a call, then in the abstract representation of that call there are two connections, and each connection corresponds to one of the devices.

As we shall see in Chapters 4 and 6, connections are the most used element of this abstraction because they allow efficient manipulation of both a call and a device simultaneously.

3.2.6. Graphical Notation

The abstraction of calls, devices, and connections is illustrated graphically using a notation based on simple symbols.

Calls are represented as circles as shown in Figure 3-5. In order to make reference to a specific call, the calls are labeled with the letter "C" followed by a number.

Figure 3-5. Symbol for calls

Devices are represented as rectangles, as shown in Figure 3-6. They generally are labeled with the letter "D" followed by a number, so that different devices can be explicitly discussed.

Figure 3-6. Symbol for devices

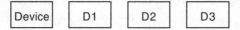

Connections represent the relationship between a call and a device. They are represented graphically as a line between a device and call, as shown in Figure 3-7. Connections do not require explicit labels because they can be uniquely identified by making reference to the labels for the device and the call (in that order) that they associate. Figure 3-7 shows two devices, D1 and D2, connected to call C1 with connections D1C1 and D2C1 respectively.

Figure 3-7. Symbolic representation of connections

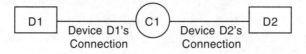

Figures 3-8, 3-9, and 3-10 illustrate the use of this notation in a number of typical examples. In the first example, call C1 has two connections D1C1 and D3C1. The second example shows a multi-device call in which D1, D2, and D3 are all participating in call C2 using connections D1C2, D2C2, and D3C2 respectively. The last example shows call C3 in the process of being set up or cleared, as it is associated only with connection D4C3.

Figure 3-8. Two-device call

Figure 3-9. Three-way call

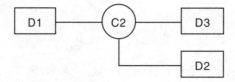

Figure 3-10. Single device in a call

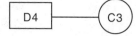

The definition of this notation will be expanded in the next few sections.

3.3. Switching

Switching refers to the function that establishes and clears calls and manipulates the connections associated with those calls. Switching is therefore the most fundamental and essential activity that takes place in a telephone system. The telephony resources that implement switching in a system are referred to as *switching resources.*

Switching resources manage calls by:

1. Appropriately tracking and conveying control information associated with a call, and

2. Conveying the media stream associated with a connection by allocating and deallocating channels.

The operations that are performed by switching resources are referred to as *switching services.*

3.3.1. Media Stream Channels

Media stream channels are paths of communication that can be established to convey a media stream. There are a finite number of media stream channels in any implementation of switching resources, and a finite number of media stream channels that can be associated with connections for any given device. A media stream channel is able to carry a single media stream in each direction simultaneously.

When a channel is in use it is referred to as *allocated,* and when it is retired from use it is said to have been *deallocated.*

3.3.2. Voice and Digital Data: Quality of Service

By default, references to connections represent media streams that are compatible with the *voice network*. This means that they carry voice or modulated data (see the sidebar "Modulated Data"). A *voice call* is therefore a call made up of *voice connections*. A rule of thumb is that if a piece of analog telephony equipment can be added as a device on the call, the call is considered a voice call.[3-3] A voice connection allocates, at most, a single media stream channel.

Some switching implementations are able to create connections that are associated with digital data media streams. *Digital data media streams* support data traveling at much higher rates than are possible with modulated data because it takes advantage of the digital switching capability in a *digital data network*. These *digital data calls* (calls made up of digital data connections) are treated specially by the switching implementation, however, because they cannot interoperate directly with voice calls (calls made up of voice connections) and only a subset of switching functionality applies to them. A digital data connection may be associated with any number of media stream channels.

The nature of a call or connection (i.e., whether it is voice or digital data) and attributes relating to digital data rates, the number of media stream channels used, and transmission characteristics, are referred to as *quality of service* or *QoS*. Quality of service is a property that is common to a call and all connections in that call. The quality of service applicable to a new call or connection is limited by what a given switching implementation supports. The quality of service associated with an existing call or connection determines the switching services that may be applied to it.

3-3. Voice call — Technically speaking, a *voice call* is one that carries media streams requiring no more than 3.1 kHz of bandwidth.

Modulated Data

Telecommunications (the transmission of digital data by computers) is accomplished either by connecting a computer directly to the digital data portion of the telephone network, or by transmitting the data over the voice network using modulation technology.

Modulation involves transmitting a *carrier* signal, that is, a tone of a particular frequency, and then varying it according to the rules of a particular modulation protocol. Modulation protocols dictate the use of techniques that vary the carrier signal amplitude, frequency, phase, or other properties, often in combination. *Simplex* data transmission occurs when there is only one carrier signal on the media stream, so data transmission can take place only in one direction. The term *half-duplex* means that the devices at either end of a simplex media stream are able to take turns transmitting and receiving. *Full-duplex* communication takes place when two different carrier frequencies are used on the same media stream so that data can be both transmitted and received simultaneously. *Demodulation* involves decoding the modulated carrier signal into the original digital data stream.

A modulator-demodulator, better known as a *modem*, is a piece of hardware or software that is able to perform the modulation and demodulation of signals, given full access to a media stream. Hardware-based modems are typically packaged along with all of the electronics needed to establish and gain access to the media stream on a call. Software modems, on the other hand, depend on independent hardware and software mechanisms to deliver the required isochronous media stream data.

The modulation protocols and other related standards that allow modems from various vendors to interoperate are established by the ITU-T and are assigned names beginning with "V." Commonly used V-series standards include V.32, V.32bis, and V.34, which define full-duplex 9600 bps, 14400 bps, and 28800 bps protocols, respectively. The V.42 standard defines error detection used with V.32, and the V.42 bis standard defines the compression algorithm for use with V.42. The use of compression allows data transmission rates that are much higher than the raw data rates of the modulation protocols. The V.17 standard defines the modulation protocol used for the exchange of documents between Group 3 fax products.

3.3.3. Directional Streams

By default, all connections and their associated media stream channels are *bidirectional*; that is, there is a media stream associated with the call that is flowing to the device, and another media stream flowing away from the device.

Some switching implementations are able to create unidirectional connections. In these *unidirectional connections* there is only one media stream, and it flows either toward or away from the device. Figure 3-11 shows how unidirectional connections are represented in graphical notation.

Figure 3-11. Directional connections

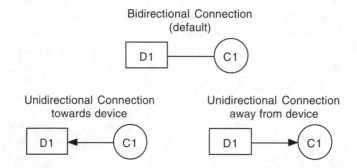

All connections and associated media stream channels in a particular call also are symmetric. *Symmetric connections* involve media streams that deliver data at the same rate in both directions. In an ordinary telephone call, this means that the same amount of sound information is being sent in both directions.

Asymmetric communication, where the quality of service is different in one direction relative to the other is abstracted as two separate calls consisting of unidirectional connections. In Figure 3-12, call C1 could be a high-speed data call, while call C2 could be a low-speed data call.

Figure 3-12. Asymmetric communication

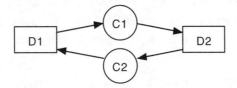

3.3.4. Point-to-Point and Multi-Point Calls

Most calls are point-to-point calls. This means that the call involves associating two devices with two connections. However, most switching implementations support calls with three or more connections. These are referred to as *multi-point calls*.

In the basic case of a multi-point call consisting of bidirectional voice connections, all of the media streams coming from the associated devices are combined and the result is sent back out to all the devices.[3-4] The effect is that all devices "hear" everything on the call.

Figure 3-13. Point-to-point and multi-point calls

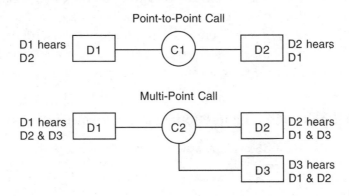

Multi-point calls are applicable to voice calls only. Audio is a unique data type in that in its analog representation it may simply be mixed with other audio sources and the result is still a useful audio stream.

3-4. Conferencing — Some implementations may use signal processing capabilities to eliminate echo effects and send a slightly different media stream to each device.

The same cannot be done with digital data streams or modulated data streams. Adding these data streams together would simply corrupt the data. Synthesizing multi-point calls for digital data and large numbers of voice calls is described in the sidebar "Synthesizing Multi-Point Calls."

Of course, every connection in a multi-point call need not be bidirectional. A multi-point call can be made of any combination of bidirectional and unidirectional connections as long as the media streams are all voice. Three special cases of unidirectional connection usage in a multi-point call are discussed below.

A multi-point call is generally referred to as a *conference call*, although the term *bridged call* is also used when the implementation is hard-wired in some way.

Silent Participation: Tapped Calls

One special case of a multi-point call is referred to as *silent participation*. In this case, a unidirectional connection is added to a call such that the new device can hear everything on the call but cannot be heard by the others participating in the call. This also is commonly referred to as a *tapped call*. Silent participation is very useful in situations where media access resources are to be connected to a call for the purpose of recording audio or doing speech recognition, for example.

Figure 3-14. Multi-point with unidirectional stream towards device

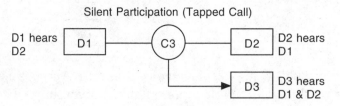

Silent Participation (Tapped Call)

D1 hears D2 — D1 — C3 — D2 — D2 hears D1

D3 hears D1 & D2 — D3

Synthesizing Multi-Point Calls

A multi-point call is one in which switching resources take the media streams from all connections to the call, mix them together, and deliver the result to all of the participants that are receiving a media stream from the call.

In situations where the switching resources cannot combine two or more calls into a multi-point call (because they are digital data calls, for example), multi-point calls can be synthesized using a series of point-to-point calls in a centralized or distributed fashion.

In the centralized approach, all of the participating devices establish point-to-point calls to a *conference bridge*, which is a special piece of equipment that is capable of decoding or interpreting the media stream sent from every participating device and intelligently mixing the media streams together. Each participating device only sees a single other device in the point-to-point call it establishes to the conference bridge, but it receives a media stream specifically tailored to be the device's "view" of all of the other devices. A conference bridge may handle just voice data (good for cutting down the noise on multi-point calls with many people), or may handle specific protocols using digital data, such as videoconferencing calls. In fact, a sufficiently intelligent conference bridge (also called a *multi-point control unit* or *MCU* in the context of H.320 videoconferencing) can mix caller types, so that a voice call to a conference bridge on which a videoconference was being hosted would hear a special mix of the audio from the videoconference but wouldn't see the video.

The distributed approach to synthesizing multi-point calls involves each participating device placing a call to every other device. As with the conference bridge, the transmitters prepare a specific version of the media stream for each participant to which they are connected. In the distributed case, however, it is the responsibility of each participant to do the appropriate "mixing" of the media streams being received so they can be presented as a single media stream to the person actually using the device in question.

Announcement

Another special case of multi-point calls is the *announcement*. As shown in Figure 3-15, this involves the addition of a unidirectional connection to the call that delivers a media stream from the added device. This is used primarily for adding a media access resource that plays an announcement.

Figure 3-15. Multi-point with unidirectional stream away from device

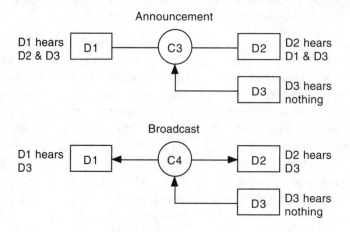

Broadcast

A third special case of multi-point calls is the *broadcast*. In this case, all of the connections in a call are unidirectional. Exactly one of the connections is unidirectional away from its device and all the others are unidirectional toward the devices. This allows one media stream to be delivered to many devices.

States and State Diagrams

The term *state* refers to the mode or condition of some entity. An entity's state indicates settings of its attributes and its current circumstances, thus determining what it can and cannot do. A given entity can be in only one state at any given instant.

A *state diagram,* or *state graph*, represents the states of which an entity is capable, as well as the transitions between states that are permitted. Once an entity is in a given state, it can only assume a new state if the state transition model represented in the state diagram indicates that the change is allowed.

In the example below, an entity is permitted to transition between three states identified as *one, two,* and *three.* According to the graph, the entity can transition from state *three* to state *one*, from state *one* to state *two*, and may go back and forth between states *two* and *three*. It may also transition from state *two* back to state *two.*

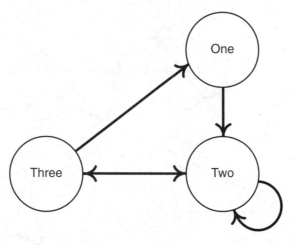

States greatly simplify the modeling of systems because, once an entity is in a known state, there is only a finite set of states to which it can transition.

3.3.5. Connection States

Connection states are among the most important concepts in telephony. Each and every connection has an individual state that determines the condition of the connection and what can and cannot be done with that connection. The life cycle of a device's participation in a call (or call progress) is represented by the sequence of states through which the corresponding connection transitions. A state transition indicates that some service was performed on the connection by the switching implementation. The switching services that may be applied to a connection at any time are determined by its connection state.

Figure 3-16. Connection state diagram

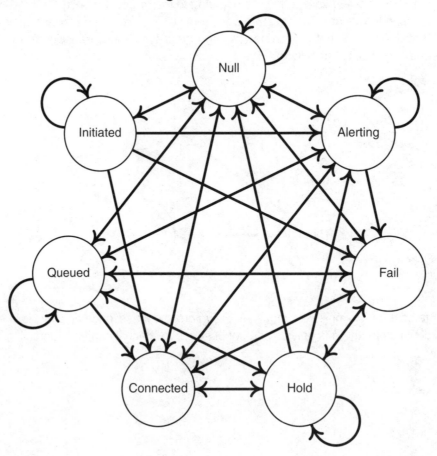

There are seven connection states (described below), and they are permitted to transition as shown in the state diagram in Figure 3-16.

Initiated

While a device is requesting service or dialing a digit sequence to initiate a call, the corresponding connection is in the *initiated* state. The *initiated* state typically is entered when the device goes off-hook[3-5] and, if supported, when the user is prompted[3-6] to take the device off-hook. The media stream received by the device, if any, is initially dial tone followed by silence after the first digit or command is issued.

Connected

After a call has been created and the telephone system is establishing connections with other devices, the connections where media stream channels are allocated and the associated media stream(s) are flowing, are in the *connected* state.

Null

A connection is said to be in the *null* state if it no longer exists. A nonexistent connection is synonymous with a connection in the *null* state. Occasionally a connection is said to "transition through the *null* state." This is a simple way of saying that the original association between a device and call was replaced with a new one.

3-5. Off-hook — In general usage, the term *off-hook* refers to an action to be taken on a device, typically lifting a telephone handset, signifying that the device is requesting service. The term comes from the days when a telephone's microphone hung from a hook. The use of the term off-hook as it applies to the operation of analog telephone sets is explained in Chapter 5.

3-6. Prompting — The term *prompting* refers to an indication, typically audible, a telephone set makes to indicate that it should be taken off-hook in order to progress out of the *initiated* state. Prompting is distinct from *ringing*, which is an indication made by a device that a connection is in the *ringing* mode of the *alerting* state. Prompting is used when some service is initiated for a device but is unable to go off-hook without manual intervention.

Alerting

While an attempt is being made to connect a call to a device, the connection representing the association between the call and device is in the *alerting* state. There are actually three *modes* of the *alerting* state that determine what type of action the device can take:

- *Entering distribution* mode

- *Offered* mode

- *Ringing* mode

These modes will be discussed in Chapter 4.

The fact that a connection is in the *alerting* state is independent of whether the corresponding device is indicating an incoming call in some way (by ringing, for example).

Typically for voice calls, all of the other active (e.g., *connected*) connections to the call will hear ringback,[3-7] while a connection in the *alerting* state is associated with the call.

Fail

The *fail* state indicates that call progress was stalled for some reason and an attempt to associate a device and a call (or keep them associated) failed. The most common example of this state is attempting to connect to a device that is busy.

In most cases, all of the other active (e.g., *connected*) connections to the call will hear a busy tone, or another appropriate failure tone, while a connection in the *fail* state is associated with the call.[3-8]

3-7. Ringback — *Ringback* is the "ringing" sound heard after you have placed a call and are waiting for it to be answered. It is not the actual sound of a phone ringing, but rather a sound the switching implementation generates to supply the caller with feedback on call progress.

3-8. Blocked — One case where the *fail* state may not be associated with an audible tone is the case where a bridged connection is blocked from a call. This is described in section 3.9.3.

Hold

When a connection is in the *hold* state, it continues to associate a particular device with a call (signaling information continues), but the transmission of associated media streams is suspended. Depending on the implementation, the media stream channel(s) associated with the connection's media stream(s) may or may not be deallocated while the connection is in the *hold* state. A channel that is not deallocated is said to be *reserved*.

Hold should not be confused with mute. Mute is a telephone set feature which will be described later in this chapter. Mute deals with turning off a speaker or microphone and is not related to the transmission of media streams or the allocation of media stream channels.

Queued

A connection is in the *queued* state when call progress is suspended pending subsequent application of certain switching services. Like the *hold* state, connections in the *queued* state do not have active media streams and the associated channel may or may not be deallocated.

3.3.6. Connection State Representation

In graphical notation, connection states are represented by placing symbols over the line representing the appropriate connection. The symbols for the states are:

- 'a' represents *alerting*
- 'c' represents *connected*
- 'f' represents *fail*
- 'h' represents *hold*
- 'i' represents *initiated*
- 'n' represents *null*
- 'q' represents *queued*

If a particular connection is in the *null* state, meaning that it doesn't actually exist, the line representing the connection is generally omitted from the diagram altogether.

Figure 3-17 shows an example of how these symbols are used in graphical notation.

Figure 3-17. Connection state representation

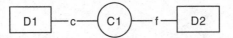

In this example, call C1 has two connections: D1C1 and D2C1. Connection D1C1 is in the *connected* state, indicating that it is active and that media streams are flowing. Connection D1C2 is in the *fail* state. A likely reason for the situation depicted in the example is that device D1 tried to place a call to device D2 but D2 was busy, so the attempt to connect to D2 failed. In this case, Device D1 probably would be receiving busy tone from the call.

3.3.7. Switching Services

Switching resources carry out operations on the calls and connections in a telephone system. In terms of the basic telephony abstraction, these *switching services* involve one, some, or all of the following five actions:

1. Creating new calls

2. Disposing of calls

3. Adding connections to a call

4. Removing connections from a call

5. Manipulating connection states

Each switching service changes the relationships between particular devices and calls by acting on connections that relate them.

A given switching service therefore can be described in terms of the states of the connections to which it can be applied, the state transitions that result, and whether it creates or disposes of calls or connections. This is generally presented in terms of "before" and "after" the switching service has taken place.[3-9] A switching service is represented as shown in Figure 3-18.

Figure 3-18. Switching service representation in terms of "before" and "after"

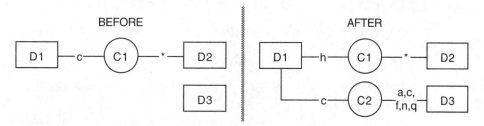

In this example, the switching service shown is "Consultation Call" being applied to the connection D1C1. Before the service, the connection D1C1 is in the *connected* state. After the service, D1C1 is in the *hold* state; a new call, C2, has been created; and the connection D1C2 is in the *connected* state.

If a connection might be in one of a number of different states before or after a switching service, this is shown in the notation either by providing a list of the symbols (as shown in the preceding example), or by using one of the following special symbols. The meaning of one of these symbols depends on whether it is on the "before" or "after" side of the diagram.

3-9. Call control services — Switching operations with well defined state transitions and other behavior are referred to as *call control services*. Refer to Chapter 4 for details of specific call control services. Between the states shown as "before" and "after" for a given service, connections may transition through intermediate states. Each complete sequence of event transitions is referred to as a flow. The concepts of flows and normalized flows are explained in Chapter 6.

- '!' represents *unspecified*

    - This symbol is equivalent to "a,c,f,h,i,n,q"

    - Before a service, it indicates that any connection state is applicable.

    - After a service, it indicates that any connection state may result and the original state (if the connection existed before the service) has no bearing on the final state.

- '#' represents *unspecified non-null*

    - This is equivalent to "a,c,f,h,i,q"

    - Before a service, it indicates that any connection state other than *null* is applicable.

    - After a service, it indicates that any connection state other than *null* may result and the original state (if the connection existed before the service) has no bearing on the final state.

- '*' represents *unspecified/unaffected*

    - Before a service, it indicates that any connection state is applicable and that the connection state will be unaffected by the service.

    - After a service, it indicates that the connection state was not affected by the service and will be the same as it was prior to the service.

Figure 3-19. Representation of the Clear Connection[3-10] switching service (applied to D2C1)

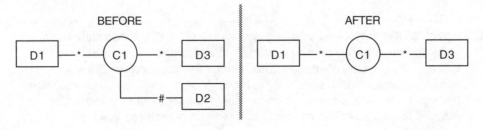

An example using these symbols is shown in Figure 3-19. In this example, the switching service *Clear Connection*[3-10] is applied to the connection D2C1. Before the service, there are three connections to the call (D1C1, D2C1, D3C1) and the only restriction is that D2C1 be in any state other than *null*. After the service, there are only two connections (C2D1, C2D3). The connection states for D1C2 and D3C2 were not affected by the service.

The set of switching services supported by a particular product represents a large portion of its telephony feature set. See Chapter 4 for descriptions of all the common switching services and other telephony features.

3.4. Telephone Networks: Behind the Cloud

In Figures 3-2, 3-3, and 3-4, a cloud was used to represent the telephone network in which a call existed. These clouds[3-11] represent the switching fabric, or the set of telephony resources and the connections between them, being used to establish a call.

3.4.1. Switched Networks

To illustrate the rationale behind switching, imagine that you needed a separate telephone for every person you called, and that each telephone was connected directly to a corresponding telephone owned by each of the other people. Not only would such a system be quite

---

3-10. Clear Connection example — Note that this is actually just one case of the *clear connection* operation, where D2 is not a shared–bridged device (see section 3.8.3) and both D1C1 and D3C1 are non-*null*.

3-11. Clouds — The cloud symbol has become the standard graphical representation of a switching fabric or network. It is a convenient abstraction because one need not be concerned with how large or small the network is, or what portion of the resources it represents are consumed in taking care of a given call. If the network is operational, all the endpoints on a call are appropriately connected. Without this ability to work with a simple abstraction of the overall network, even the simplest operations would be quite complicated to explain. It has been said that telephone companies "are in the cloud business."

impractical, it also would be very inefficient because you can use only one telephone at a time. An example of this is shown in Figure 3-20. A seven-endpoint fully connected network (where every endpoint has a direct channel to every other) contains 21 direct media stream channels, yet only three of these can be active at a given time (assuming that each endpoint can only use one channel at a time). All of the capacity represented by the other dedicated media stream channels and telephones would go unused most of the time. In contrast, a switched network of the same size has only seven media stream channels, and all of these can be utilized simultaneously (assuming the switching resource support multi-point calls). This is why the notion of switching was developed.

Figure 3-20. Networks

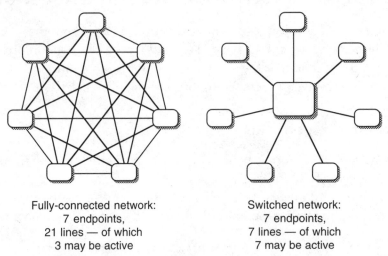

Fully-connected network:
7 endpoints,
21 lines — of which
3 may be active

Switched network:
7 endpoints,
7 lines — of which
7 may be active

Switching allows a limited number of media stream channels, lines,[3-12] or *transmission facilities* to be used as efficiently as possible by sharing them. This is referred to as *line consolidation*. Telephone networks are

3-12. Lines — The terms *line* and *telephone line* can be misleading because many types of telephone lines are capable of supporting multiple channels. While the term line tends to imply a piece of wire, it actually refers to any type of link. The concepts shown here apply equally to wireless networks and fiber-optic networks. Chapter 5 explores different types of telephone lines.

systems of these facilities, interconnected with switching resources that ensure the calls always reach their destinations and a minimum number of media stream channels are needed to connect any two points. This is particularly important as the size of the network increases. Figure 3-21 shows the interconnection of two small networks. Here only two additional media stream channels are needed two connect the two networks into a larger network, by directly connecting the switching resources. This ability to scale is theoretically unbounded; for example, the endpoints shown in the diagram could themselves actually represent different networks.

Figure 3-21. Joining two switched networks

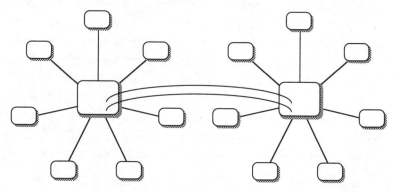

In the particular example shown, it was determined that most calls established by the members of each small network are *internal* to those networks, and that the need for more than two active calls between the networks is extremely rare. This provides insight into the limitation of switched networks: that the ability to establish calls between two points may be limited by the capacity of the transmission facilities between two sets of switching resources somewhere in the network. In practice, the allocation of transmission facilities based on usage statistics represents a very important science in the business of operating a switched network. If the estimate for capacity requirements is incorrect for a specific span between two adjacent nodes, the network may have underutilized capacity, may consume more switching resources than needed in order to reroute calls around overloaded spans, or may simply fail to complete calls if no paths exist.

3.4.2. Public, Private, and Virtual Private Networks

There are a number of types of telephone networks. The primary difference between these is related to ownership or allocation of the telephony resources and transmission facilities.

- Public networks

    The term *Public Switched Telephone Network* (*PSTN*) refers to the publicly accessible worldwide telephone network that ultimately interconnects with virtually every other telephone network. The public network is made up of many individual public networks around the world. Any network within the worldwide network (France Telecom's network, for example) is itself a public network. Some portions of the public network are owned by commercial telephone companies and others by government-run enterprises. Collectively these service providers are referred to as *common carriers*. In all cases, the operators of the public network are intensely regulated by regulatory bodies in each country. In general, this regulatory effort is aimed at ensuring universal access to service and fair or, if appropriate, competitive rates.

- Private networks

    Private networks are sets of telephony resources that are privately owned or leased from a common carrier and connected together with dedicated transmission facilities. A private network can be as simple as a home's internal phone system. It can be as complex as a worldwide network made up of switching resources in hundreds of locations that interconnect with the public network in many different countries. Private networks may be operated for the exclusive use of their owners, or may be operated by private carriers who sell access to their networks, or some combination of the two.

- Virtual private networks (VPN)

    A *virtual private network* is a capability provided by a common carrier that offers service equivalent to a private network but uses the shared facilities of the public network. The result generally is a more cost-effective network because unused capacity is not wasted.

3.4.3. Multiple Carriers in the Public Network

In many parts of the world, multiple common carriers compete for customers. In these environments, customers may choose which carriers operating in the public telephone network they wish to have carry their calls. The ways that the networks of competing carriers can be accessed varies from one location to another and from one carrier to another.

Depending on the type of carrier and the context, any combination of the following arrangements are possible:

- Direct access

    In most cases a private telephone system is connected physically to a single specific carrier, referred to as the *local exchange carrier* (*LEC*) or *dialtone provider.* In the case of a wireless telephone, the actual carrier used is determined dynamically. A private telephone system with more sophisticated switching resources can be connected directly to two or more carriers (Figure 3-22). In this case, the telephone system decides which carrier to use for a given call. In the case of a *carrier bypass* arrangement, one of these carriers provides local service and one provides long-distance service. It is called carrier bypass because access to the long-distance carrier bypasses the local carrier's network. This also may be called a *foreign exchange* (*FX*) facility if the second network is actually a local network in a different location.

Figure 3-22. Directly connected carriers

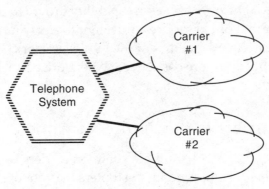

- Default carrier

    A local exchange carrier may be used to access alternative carriers where a default carrier is established for specific types of calls (Figure 3-23). For example, one could specify that all long-distance calls are to be routed through a specific long-distance *interexchange carrier*, or *IXC*.

Figure 3-23. Default carriers

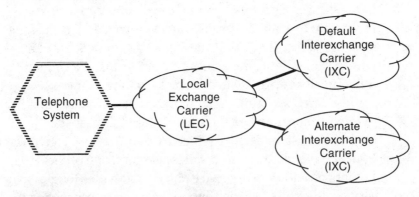

- Dialed carrier

    With each outbound call, the local exchange carrier is informed which alternate carrier should be used. This typically is accomplished by dialing special sequences when establishing calls (Figure 3-24).

Figure 3-24. Dial selectable carriers

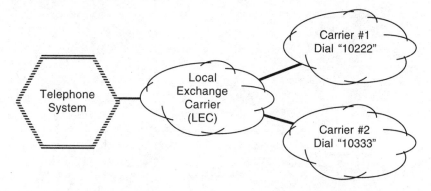

3.4.4. Telephony Resources in a Telephone Network

A telephone network is made up of a "web" of interconnected sets of telephony resources. To establish a call between two devices, one or more sets of telephony resources are used to patch together a complete end-to-end call across the network. Figure 3-25 shows a call that has been established between device D1 and device D2 across a telephone network. The network "cloud" is used to abstract the network; the call between the two devices then is easily abstracted.

Figure 3-25. Call between D1 and D2

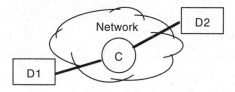

Figure 3-26 shows the same scenario, but instead of abstracting the network as a cloud, we see the actual system of individual sets of telephony resources (represented as hexagons) that make up the network. The call between D1 and D2 is actually a series of calls. Each participating set of telephony resources establishes a different segment of the overall path that the call takes through the network. Each segment of the complete call is actually an individual call made up of connections with two devices, just as in the overall abstraction.

Figure 3-26. The network of telephony resources behind the cloud

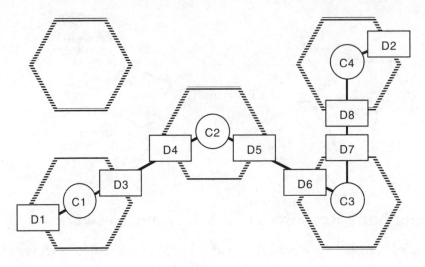

3.4.5. Network Interface Devices

Figure 3-26 illustrates another important concept: the *network interface device*.

Up to this point we have been using the term *device* to refer to a general class of telephony resources capable of having a connection to a call. The first example of a device was the telephone set, which is known as a station type of device. Station devices[3-13] actually terminate or originate media streams and can be thought of as being *local*, or directly attached, to a particular set of telephony resources.

3-13. Device configurations — Devices are further subdivided into elements that are arranged according to a device configuration for the individual device. There are two types of device elements, *physical elements* and *logical elements*. Unless otherwise stated, all references to devices are to logical device elements. Logical device elements and device configurations are explained in sections 3.8 and 3.9. Physical device elements are explained in section 3.6.

A *network interface device* is an endpoint with respect to one set of telephony resources, but corresponds to a transmission facility that connects to another set of telephony resources. Figure 3-27 shows just one set of telephony resources and abstracts the rest of the network. In this case, device D3 is a network interface device.[3-14]

Figure 3-27. Network interface device

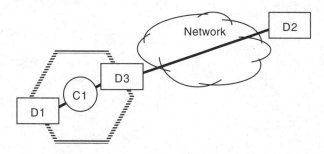

From the perspective of the telephony resources shown in Figure 3-27, the network interface device D3 is a proxy for the remote device D2 on the other side of the network. These telephony resources cannot manipulate D2, so they act on D3 in its place. For example, to drop connection D2C1 from the call, the switching resources drop connection D3C1, which has the same result. The relationship of D3 as a proxy to D2 may be represented in graphical notation as shown in Figure 3-28. The brackets indicate that D3 is a proxy for D2.

Figure 3-28. Network interface device representation

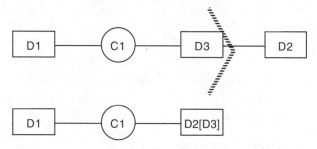

3-14. Network interface devices — Depending upon the implementation involved, network interface devices may be referred to as *trunks*, *direct lines*, or *CO lines*.

3.5. Call Routing Resources

Call routing refers to the movement of a call from device to device or, from the call's perspective, the sequence in which connections to new devices are created and cleared. In Figure 3-29 call C1 is routed first to D2, then to D3, and finally to D4.

Figure 3-29. Call routing

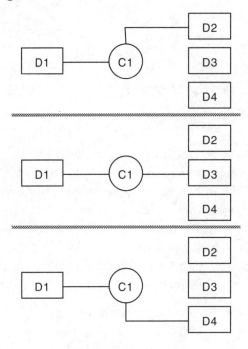

3.5.1. Call Processing

Call processing is ultimately responsible for all call routing activity as it directs the switching resources that establish and clear connections. Call processing establishes and clears calls as a result of:

- Commands from devices

- Commands from CTI interfaces

- Expiration of certain timers

- Feature settings (stored rules such as *call forwarding*[3-15] that override normal call progress)

- Default rules that determine "normal" call progress

When a device creates a call, it specifies a desired destination for the call. By default, call processing will attempt to connect the specified device to the call; the call may be redirected many times, however, before being connected. The various features and switching services that relate to call routing are explained in Chapter 4.

Among the telephony resources that may be present in a given telephone system are special devices that may be associated with a call strictly as an additional step in the routing of that call. As always, the call is associated with these special devices through connections, but the states of these connections generally are either *queued* or *alerting*.

3.5.2. Park Device

A *park device* is a special device with which calls may be associated in order to set them aside, or "park," them temporarily. Connections associating a call and a park device, if present, are in the *queued* state. A call associated with a particular park device later can be *picked* and connected to any other device in the telephone system.

3.5.3. Pick Group Device

A *pick group device* is a single special device associated with two or more devices (typically stations) that form a *pick group*. The pick group device is comparable to the park device, except that calls are not explicitly parked to it. In fact, it is a unique type of device that never interacts with calls directly. Instead, the pick group device simply keeps track of all connections associated with all of the devices in the

---

3-15. Call forwarding — Telephony features, and call forwarding in particular, are discussed in Chapter 4. Call forwarding is a feature that allows a rule to be established for the automatic redirection of a call in the event that certain criteria are satisfied.

pick group that are in the *ringing* mode of the *alerting* state, the *hold* state, or the *queued* state. Depending on the implementation of the telephone system, any device in the telephone system or just those from the pick group in question, can then request that a call be *picked* from those being tracked by a particular pick group device. Devices in the group pick their calls from the pick group by default.

3.5.4. ACD Device

An *Automatic Call Distribution* device, or *ACD* device, distributes calls that are presented to it. Depending on the implementation of the ACD, it may use any algorithm or logic for determining what device in the telephone system to redirect a call to. An ACD device is sometimes referred to as a *split*.

When a call is presented to an ACD device, a connection in the *entering distribution* mode of the *alerting* state usually is created. The ACD then specifies the device to which it wants the call redirected, and the connection to the ACD is cleared. If the ACD cannot immediately identify an available device to which to direct the call, it may either leave the call in the *alerting* state or transition it to the *queued* state until a suitable destination is identified.

An ACD may employ other devices, including media access devices (for playing messages, etc.) and other ACDs or ACD groups. Implementations of ACD devices may be modeled one of two ways. The first model involves *visible ACD-related devices.* In this model, any devices employed by the ACD for interacting with the call are modeled independently from the ACD device itself (with separate connections to the call). In the second model, *non-visible ACD-related devices*, the additional resources used are all considered to be within the ACD device itself, so only a single connection to the call is used.

3.5.5. ACD Group Device

An ACD group device is like an ACD device in that it distributes calls to other devices. Unlike the ACD device, however, it only distributes calls to a specific group of devices that are specifically associated with the ACD group device. In every other way, an ACD group device operates in the same fashion as an ACD device. An ACD group device is sometimes referred to as a split or *pilot number*.

ACD Agents

The association between an ACD group device and a device to which calls can be redirected is called an *agent*. Just as connections are dynamic entities that represent the temporary relationship between a device and a particular call, an agent represents a similar temporary relationship between a device and one or more ACD group devices.

Agents are created when a device associates itself with a particular ACD group. This is referred to as *logging onto* an ACD group. An agent is cleared after the corresponding device *logs off* the ACD group.

Just as connections have a connection state, agents have an *agent status*. An agent's status may be one of the following:

- *Agent null*

    No relationship exists between the agent and the ACD group; that is, the agent is logged off.

- *Agent logged on*

    A relationship between the device and the ACD group has been established, but no distribution of calls to the agent has yet begun.

- *Agent not ready*

    The device associated with the ACD group is not prepared to receive calls distributed by the ACD.

- *Agent ready*

    The device associated with the ACD group is prepared to receive calls distributed by the ACD.

- *Agent busy*

    The device is dealing with a call that was directed to the device by the associated ACD group.

- *Agent working after call*
  The device has completed working with a call that was directed to the device by the associated ACD group, but it is not yet prepared to deal with another call.

Agent status is used by the ACD group device for two things.

1. Routing

   The ACD group will only distribute calls to devices associated with agents that have a status of *agent ready*. It should be noted that agent status relates only to the status of the agent and not to the device itself. So a device that is busy with a call delivered by some mechanism other than through the ACD group may have a status of *agent ready*. Likewise, only the ACD groups associated with the device through the association represented by the agent are affected by the agent's status. This means that a device still may receive calls from any other devices.

2. Statistics

   Most ACD group implementations provide detailed statistics on the activity of agents. The amount of time that an agent had a given agent status, along with the times each call spent queuing before being redirected to a particular device, are generally captured so that the efficiency of the system and its users can be optimized.

3.5.6. Hunt Group Device

A *hunt group device* is like an ACD group device in that it distributes calls to a specific group of devices. Unlike the ACD group device, however, the group of devices served by a hunt group's distribution function is fixed in some fashion, so the concept of agents does not apply. In every other way it behaves in the same fashion as an ACD device or an ACD group device. Hunt group devices are also sometimes referred to as pilot numbers.

3.6. Telephone Stations

Station devices are the telephony resources that correspond to the tangible telephones and telephone lines with which we are all familiar. Station devices take an almost unlimited number of different forms. There are literally hundreds of telephone vendors around the world, each of which manufactures many different models of telephones. Each telephone is designed to appeal to the particular needs and preferences of a different type of telephone user.

Despite the fact that at the most simple level all these telephone devices provide the same basic functionality, there is an endless number of variations in the form that the physical user interface to a telephone system can take. Research has demonstrated, and the marketplace has validated, that people have very diverse and particular preferences when it comes to the form that their interface to telephony services should take.[3-16]

It is important to note that from the perspective of telephony concepts, every type of telephone, regardless of its functionality, interface to the telephone network, or other properties, is a station device. This includes POTS (plain old telephone service) telephones ("500" and "2500"[3-17] telephone sets) and multi-line telephones. It includes wireless phones such as cordless and cellular phones, and coin and card pay phones. It includes novelty phones from "football phones" through "shoe phones." It includes full-featured digital phones and attendant consoles, and hybrid devices such as fax phones and video phones.

3-16. Diversity of telephony user interfaces — The fact that user preferences are so diverse when it comes to a personalized interface to telephony functionality (traditionally the telephone) plays a very important role in both the motivation for CTI and the architecture of CTI implementations.

3-17. 500 and 2500 telephone sets — AT&T assigned the model number 500 to the old rotary-dial deskset telephones. When touchtone was developed, the touchtone model was numbered 2500. Since then, the term *2500 set* has come to be used to reference any telephone that is functionally equivalent to the original model 2500 telephone set.

3.6.1. Physical and Logical Device Elements

The portion of a device that is involved with call control, that is, the creation, management, and clearing of connections, is referred to as the *logical element* of a device. Most device types have only a logical device element.

Figure 3-30. Physical and logical elements

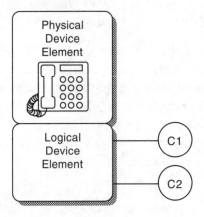

Station devices typically have a tangible portion—a physical telephone set—that is referred to as a *physical element*. To interact with calls, a physical device element may have a corresponding logical device element and/or may be associated with the logical device elements of other devices (Figure 3-30). (The concepts relating to logical elements and the relationships between physical and logical elements, referred to as device configurations, are described in sections 3.8 and 3.9.) While the logical element of a station device may be associated with both digital data and voice calls, the physical element only interacts with voice calls. In addition, a physical device element may only be associated with a single connection in the *connected* state at any given time, although any number of associated connections may be in other connection states.

Figure 3-31. Physical device element components

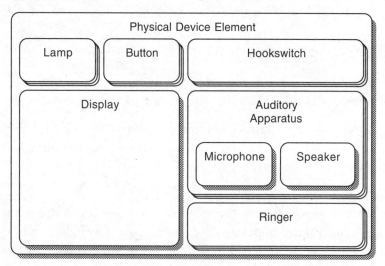

A physical device element consists of many different *components* that make up the physical user interface of the device. These components may be directly accessible to the user, or may be internal to the device (i.e., virtual). Types of physical element components include:

- Auditory apparatus
- Hookswitch
- Button
- Lamp
- Display
- Ringer

As shown in Figure 3-31, the physical element of a particular device may or may not have a display, and may have zero or more lamps, buttons, hookswitches, ringers, and auditory apparatuses. Auditory apparatuses are always associated with a particular hookswitch that governs whether they are active or not. Lamps are optionally associated with buttons.

The way that these physical device components can be combined to form different products is described in Chapter 5.

3.6.2. Auditory Apparatus

An *auditory apparatus* component is one instance of a source and destination of speech-quality media streams.

Though typically an auditory apparatus consists of both a microphone and a speaker, it need only have one or the other. An auditory apparatus usually is one of the following types:

- Handset
  A *handset* is the most common type of auditory apparatus. It is a straight or elbow-shaped device that is held in a person's hand, with a microphone at one end and a speaker at the other.

- Headset
  A *headset* is an auditory apparatus that is worn on the head. It has both a microphone and speaker. The microphone may be on a boom of some sort or built into the ear piece.

- Speaker phone
  The combination of a microphone and a speaker built into the base of a telephone set, or arranged in some other fashion to allow for untethered operation, is referred to as a *speaker phone* auditory apparatus.

- Speaker-only phone
  A *speaker-only phone* is an auditory apparatus that is like a speaker phone but has only a speaker and no microphone.

A physical element may have any number of auditory apparatuses. Every auditory apparatus is associated with exactly one hookswitch[3-18] (although a single hookswitch may be associated with more than one auditory apparatus).

Microphone

A microphone allows sound to be converted to one direction of a voice media stream. Microphones have two properties:

- Mute

  Microphone mute determines whether or not the microphone is active.

- Gain

  Microphone gain determines what amplification of the captured sound is applied.

Speaker

A speaker allows a voice media stream from the telephone system to be converted to sound. Speakers, like microphones, have two properties:

- Mute

  Speaker mute determines whether or not the speaker is active.

- Volume

  Speaker volume determines what amplification is applied to the sound.

3-18. Hookswitch devices — Because auditory apparatuses are governed by a hookswitch, they are called *hookswitch devices* by some.

3.6.3. Hookswitch

A *hookswitch* is a component that determines what auditory apparatus or set of auditory apparatuses are in use. A hookswitch can be either *on-hook* (inactive) or *off-hook* (active). If a hookswitch is off-hook, every associated auditory apparatus is actively taking in and/or putting out sound. If a hookswitch goes on-hook, every associated auditory apparatus is removed from use.

A physical element may have any number of hookswitches and each operates independently of all the others (Figure 3-32). A hookswitch may be a spring-loaded switch on the telephone, it may be a single locking switch that is pressed to set it off-hook and pressed again for on-hook, or it may be a pair of switches where one is used to go off-hook and another is used to go on-hook. It also may be an internal (or virtual) switch.

Figure 3-32. Hookswitches

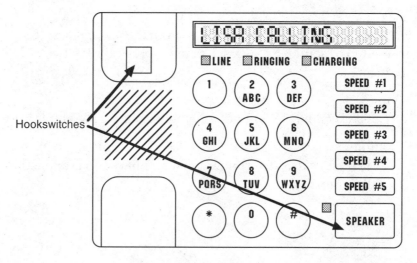

3.6.4. Dial Pad Buttons and Function Buttons

A *button* is a component that can be pressed to request a particular function or feature of the telephone system (Figure 3-33).Typically a button component is either a physical button that can be pushed and released, or an internal (virtual) switch that is pressed or triggered indirectly.

Figure 3-33. Buttons

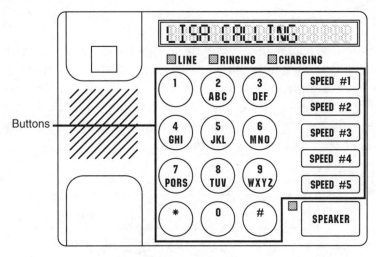

Buttons may have the following attributes:

- Label

    A button's *label* is the name by which it is called. Typically the label is printed on the surface of a physical button.

- Associated Number

    Speed-dial buttons and others that depend on remembering a particular telephone number have an *associated number*.

- Active/Inactive

    At any given time a particular button may be active or inactive. An active button can be pressed and an inactive one cannot.

Most physical element implementations have at least 12 buttons, corresponding to the buttons on a standard dial pad. Depending on the context, pressing these buttons either communicates part of a device address, indicates a particular service to be carried out, or requests that a particular tone be generated.

The most typical example of a function button is a speed-dial[3-19] button that autodials a particular number.

3.6.5. Lamps

Lamps are components that provide simple visual feedback from the telephone system. Lamps often are implemented (as the name suggests) as simple light sources on the physical element (Figure 3-34) but they can also take other forms. For example, a lamp might be implemented as a special symbol that can be made to appear and disappear on the physical element.

Figure 3-34. Lamps

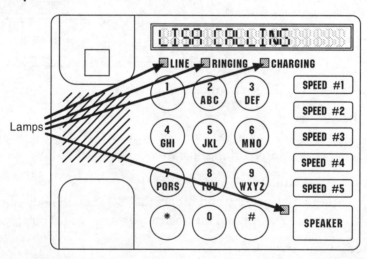

3-19. Repertory button — Speed-dial buttons are also referred to as *repertory buttons* or *rep-buttons* or just *auto-dial buttons*.

Lamps may have the following attributes:

- Label

    A lamp's *label* is the name by which it is called. Typically the label is printed on or beside the lamp. If a lamp is associated with a particular button, its label could be the same as the button's.

- Associated Button

    A lamp may or may not be associated with a single *associated button*.

- Brightness

    A lamp's *brightness* indicates the intensity of the lamp when it is on[3-20] (e.g., steady, winking, fluttering, or broken fluttering). A lamp's brightness may be one of the following:
    normal – the lamp is at its normal intensity
    dim – the lamp is dimmer than normal
    bright – the lamp is brighter than normal

- Color

    A lamp's *color* is a fixed attribute that describes the hue of the lamp when it is on[3-20]. The color may be any value but is typically one of:
    no color – meaning the lamp is just a white lamp or liquid
        crystal display (LCD) indicator
    red
    yellow
    green
    blue

3-20. "On" — Technically speaking, "on" in this context refers to the "active phase of the lamp's duty cycle."

- Mode

  A lamp's *mode* indicates what it is doing. A lamp's mode may be one of the following:

  off – the lamp is off

  steady – the lamp is on

  wink – the lamp is flashing slowly

  flutter – the lamp is flashing quickly

  broken flutter – a combination of wink and flutter[3-21]

A physical element may have any number of lamps and any number of these lamps may be associated with a particular button.

3.6.6. Message Waiting Indicator

A physical device element may or may not have a *message waiting indicator* of some sort. If present, this indicator could be implemented in a fashion similar to a lamp, could be an audible chirping mechanism, or could be implemented in some other fashion. Message waiting indicators have only two simple attributes. They are either on or off, and they are either visible or not.

3.6.7. Display

A physical device element's *display* is a grid of alphanumeric characters that allows the telephone system to communicate text-based information (Figure 3-35).

A display, if present, has the following attributes:

- Rows

  A display may have one or more *rows* of characters.

- Columns

  A display may have one or more *columns* of characters. Typically displays are 10 to 80 columns in width.

3-21. Flink — A broken flutter is referred to as a *flink* (the combination of a flutter and a wink).

Figure 3-35. Display

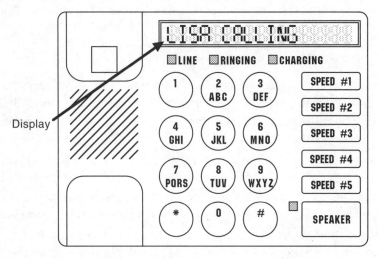

Display

• Character Set

A display's *character set* refers to the way that the data representing each character in the display, including spaces, is translated into a graphical symbol. The amount of data required to represent each character is determined by the number of different symbols supported by the display. A display's character set is typically ASCII or Unicode.[3-22]

• Contents

The *contents* of the display is the data representing the information to be displayed. The contents may be thought of as a long sequence of characters representing the concatenation of each row in the display. The size of this sequence is fixed and is determined by the number of rows

---

3-22. ASCII and Unicode — ASCII (pronounced "as-kee") stands for American Standard Code for Information Interchange. The ASCII character set uses one byte per character and can represent 256 characters of which the first 128 are standardized. The Unicode character set is designed to allow representation of the characters used by all the languages of the world, including Chinese, Japanese, and Korean. It uses two bytes per character and standardizes tens of thousands of characters.

and columns in the display and the amount of data required to represent each character (size = rows x columns x character size).

A physical device element may have, at most, one display.

The telephone system display associated with a physical device element should not be confused with other types of graphical displays that may be present on various types of products. A video phone, for example, typically is a station device with physical device elements such as buttons, an auditory apparatus, etc. Although a video phone typically has a video display that shows decompressed video data from the media stream, this display presents information from the media stream and not telephone system feedback. It therefore is not a physical element display component, although a physical element display could be implemented by superimposing it on top of the video display. In this example, the video display is modeled as a type of media access resource associated with the physical device element.

3.6.8. Ringer

The ringer is the component in the physical element that notifies people when the telephone system is attempting to connect a call to a logical device element,[3-23] with the intent that it be answered by the physical device in question (i.e., a connection in the *ringing* mode of the *alerting* state is present) or when the telephone system is prompting.[3-24]

The ringer has a bell or buzzer to provide an audible indication that the telephone set is ringing, but it also may utilize lamps or the display as *ringing indicators* that provide a visual indication that the telephone

---

3-23. Multiple logical device configurations — The relationship between physical and logical device elements is described in section 3.9. Every operational physical device element is associated with at least one logical device element, but it may be associated with two or more.

3-24. Prompting — The prompting feature is described in Chapter 4. It refers to a feature where the telephone system is unable to make a physical device element go off-hook automatically and must signal for a person to do so.

set is ringing (Figure 3-36). The ringing indicators may be built into the telephone set or may be connected to it remotely. For example, a telephone may have ringing indicators in the form of a large flashing light and bell mounted outdoors so that it can be seen and heard at a distance.

Figure 3-36. Ringing indicators

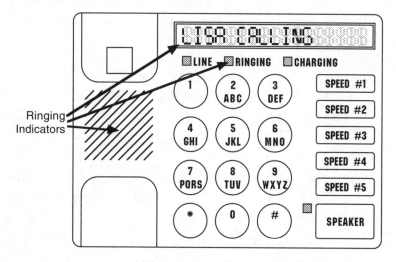

Ringers have the following attributes:

- Volume

    A ringer's *volume* is the sound level at which the ringer's audible indicator will ring.

- Mode

    A ringer's mode indicates what it is doing. It may be either *ringing* or *not ringing*.

- Pattern

    Pattern refers to the cadence, frequency, and other properties of the sound generated by a ringer when it is ringing. The pattern is set by the telephone system to indicate the type of call that is being presented, or to indicate that a particular number was called. If it is being used to indicate the type of call, it is one of the following:

unspecified – meaning call type ringing is not supported

internal – the call is from inside the telephone system

external – the call is from outside the telephone system

priority – the call has been marked as high priority

callback – the callback feature[3-25] has been used to place the call

maintenance – the call is maintenance-related (e.g., a test)

attendant – the call is from an attendant

transferred – the call is being transferred to the device

prompting – the telephone system is prompting[3-24] the device

- Count

A ringer's *count* is the number of ring cycles[3-26] that have been completed since the ringer began ringing. It is zero if the ringer's mode is *not ringing*.

A physical device element may have more than one ringer, but only one may be actively ringing at a time.

3.7. Media Access Resources

Media access resources are telephony resources, other than auditory apparatus components, that interact with the media streams associated with calls.

3-25. Call back feature — The call back feature is explained in Chapter 4, section 4.14.3. It involves a call from a device that was previously busy or unavailable.

3-26. Ring cycles — A *ring cycle* is the time between the playing of each ringer pattern while a device is ringing. The individual ringer pattern may be a simple on-off pattern or a more complex sequence. In any case, this attribute does not reflect the number or length of cycles within the pattern.

In many cases, media access resources are simply facilities within a particular device, or within the switching resources, that are transparent to the operation of that device or those resources. In other cases, media access resources are independent *media access devices* that may be associated with calls using appropriate switching services. Examples of media access devices are a generic *voice response unit* (*VRU*) or a voice mail system.

3.7.1. DTMF (Touchtone) Detectors and Generators

In existing telephone systems, the most common type of media access resources are those that generate and detect DTMF tones.

DTMF generators typically are built into most station devices so that people can generate touchtones by pressing dial pad buttons. In some telephone systems, DTMF tones are generated only to communicate a request to the other endpoints in the call. In other systems, the DTMF tones also are used to send commands (such as the number of a device to be called) to the call processing resources. In this case, general-purpose *DTMF detectors* are required by the telephone system to interpret the commands issued by various devices.

3.7.2. Pulse Detectors and Generators

Before the development of the DTMF scheme for encoding digits, *pulse* sequences were used to communicate commands. A pulse is a very short break (less than 0.1 seconds) in the media stream. A sequence of pulses represents the corresponding number, and each number is separated by a pause of at least 0.7 seconds. Pulses are typically generated by a rotary dial on a telephone station. *Pulse detectors* and *pulse generators* are telephony resources responsible for working with pulse encoded digits.

DTMF

DTMF stands for *Dual Tone Multi Frequency*. It refers to a standard mechanism for encoding the 16 digits that can appear on a telephone dial pad as combination of tones that are easily generated and detected in the media stream of a voice call. DTMF tones are also frequently referred to as *touchtones*.

As the name implies, DTMF tones are formed by unique combinations of two precisely defined tones. The scheme uses a set of four "high" tones and a set of four "low" tones. Using a combination of one low and one high tone (for 16 combinations) makes the detection of digits much more reliable; it is unlikely that some other source of media information on the call (like a person speaking) would accidentally generate one of these precise combinations of tones.

Generating these tones is also very easy to implement through a push-button interface, as shown below. Each column is associated with one of the four high-frequency tones and each of the rows is associated with one of the four low-frequency tones. Pressing a particular button connects tone generators for the appropriate row and column to the call in order to make the appropriate dual-tone or "touchtone."

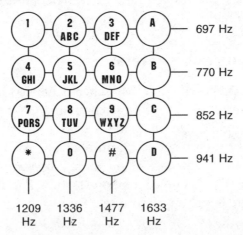

The first three columns represent the twelve DTMF tones in widespread use. The last column represents four additional DTMF tones labeled "A" through "D" which are generally referred to as *military* of *autovon* tones. These additional tones should not be confused with the alphabetic labeling found on most telephone dial pads.

3.7.3. Telephony Tone Detectors and Generators

While various mechanisms exist for providing feedback from a telephone system to the person using a particular station device, the only lowest common denominator is the media stream itself.

As a result, telephone systems working with the voice network use various tones that can be generated and detected easily, in order to provide feedback and to indicate that modulated data is to be used. These are referred to as *advisory tones* and *telecom tones* respectively.

The following are the tones that are most commonly encountered in a telephone network:

- Dial Tone

    Dial tone is an advisory tone indicating that call processing has created a new call on behalf of a given device and a new command can be sent. It is typically associated with the *initiated* connection state.

- Billing

    A *billing tone* is an advisory tone indicating that call processing is expecting to receive billing information, typically a credit card or calling card number. This tone is also referred to by some as *bong tone*.

- Busy

    Busy tone is an advisory tone indicating that there is no device currently available to which to present the call. Call progress for the call in question has stalled, so this tone is associated with a connection state of *fail*.

- Reorder

    Reorder tone is an advisory tone indicating that a call has become blocked in a telephone system because a necessary network interface device or other switching facility is unavailable. The cause may be a misdialed number. This tone is also referred to by some as *fast-busy*. Call progress for the call in question has stalled, so this tone is associated with a connection state of *fail*.

123

- Special Information (SIT) Tones

  Special information tones, or *SIT tones*, are sequences of three precisely defined tones used to indicate that call progress has stalled for some specific reason. SIT tones usually precede a prerecorded message describing the problem. They are associated with a connection state of *fail*. The four SIT tones currently in use are:

  Vacant code – the number dialed is not assigned.

  Intercept – all calls to the number dialed are being intercepted (typically because the number has changed).

  No circuit – no circuits are available for the call.

  Reorder – the call cannot be placed; the number may have been misdialed.

- Ringback

  Ringback is an advisory tone indicating that call processing is attempting to connect the call to another device. Ringback is the "ringing" sound that a caller hears after placing a call and while waiting for it to be answered. It is associated with the *ringing* mode of the *alerting* connection state.

- Beep

  A *beep* is the tone generated by a voice answering machine, voice mail system, or other media access device that records from a voice media stream. The beep is a prompt to the person calling, indicating that the recording has begun and that he or she should begin speaking (e.g., "...please leave your message after the beep...").

- Record Warning

  A *record warning* tone is a short 1400 Hz tone used to indicate that a conversation is being recorded. It is required by law in some places and typically is used in all situations (such as 911 services and security dispatch) where there is a high degree of accountability or the need to collect evidence of a conversation through a call.

- Fax CNG

    Fax CNG, or fax calling tone, is a tone generated by a fax machine or fax modem that wishes to initiate data modulation for fax transmission on a given voice call. It is a telecom tone.

- Modem CNG

    Modem CNG, or modem calling tone, is a tone generated by a modem that wishes to initiate data modulation on a given voice call. It is a telecom tone.

- Carrier

    The detection of *carrier* refers to the presence of modulated data transmission on a given voice call. It is a telecom tone.

- Silence

    Silence is the absence of any tones, voice, or modulated data in the media stream associated with a voice connection. It is quite useful to have a silence detection capability to determine that a modem transmission has completed, that a caller may have hung up, or that a message may be played.

The actual frequencies and cadences corresponding to a tone of a particular meaning may vary widely from country to country or between implementations of telephony products. This makes implementing guaranteed detection of these tones in every case very difficult, if not impossible. There is sufficient standardization, however, that tone detection for calls within a particular country or system can be reliable.

3.7.4. Media Services

Media service interfaces provide external access to the contents of media streams using specific capabilities of individual media access resources known as *media service instances*.

Some media services capture information from the media stream. They are able to transform the media stream into an appropriate data format for use with the media access resource. Other services are able to transform data received from the media service interface into media data to be placed into the media stream. Still others are able to do both simultaneously.

The number of possible types of media services is virtually unbounded, but the most popular involve working with raw sound data, speech data, modulated data, and digital data.

Live Sound Capture (Isochronous)

A *live sound capture* media service is able to capture the raw sound from a media stream and deliver it in an isochronous fashion. The media interface involved might be digital or analog.

Live Sound Transmit (Isochronous)

A *live sound transmit* media service is able to transmit an isochronous stream of raw sound delivered from a media interface to a media stream. The media interface involved might be digital or analog.

Sound Record

A *sound record* media service is able to capture sound from the media stream and store it for future use. In this case, the media service interface is used simply to start and stop the recording and specify where and how the sound is to be stored.

Sound record is different from sound capture in that the telephone system itself is doing the recording, and the sound data never leaves the telephone system.

Sound Playback

A *sound playback* media service is able to play previously recorded sounds to the media stream. In this case, the media service interface simply is used to start and stop the recording and specify what sound is to be played.

Sound playback is different from sound transmit in that the telephone system itself is what provides the prerecorded sound.

Text-to-Speech

A *text-to-speech*, or *speech synthesis*, media service is able to transform text into a stream of speechlike sounds generated by a synthetic, electronic voice. The media service interface is used to specify the text to speak and the attributes (male/female voice, accent, prosity, volume, speed, etc.) of the speech desired.

Text-to-speech is very useful because it allows arbitrary or dynamic text information to be spoken over the phone automatically. The alternative, to prerecord all of the necessary information or, at a minimum, all of the necessary words that make up the information, is generally much more complicated and expensive.

Concatenated Speech

Concatenated speech is a media service comparable to speech synthesis, but it uses strings of whole prerecorded words or syllables rather than synthesizing each syllable. Concatenated speech generally provides much higher quality than text-to-speech, but is limited to a certain vocabulary of prerecorded words or sounds.

Speaker Recognition

Speaker recognition media services identify the person speaking in the media stream, based on voice energy characteristics unique to each individual.

Speech Recognition

Speech recognition services convert human speech in a media stream to text. The principal attributes of a speech recognition implementation are:

- Speaker-dependent/independent
    Some speech recognition implementations must be trained to understand the speech patterns of a particular individual. These are called *speaker-dependent* systems. Other

implementations have been extensively trained and can understand virtually any speaker of a given dialect. These are called *speaker-independent* implementations.

- Continuous/Discrete
  Continuous speech recognition implementations are those that can automatically identify word breaks, allowing speakers to talk continuously (normally). *Discrete* speech recognition implementations cannot identify word breaks. A person speaking to such a system must place distinct pauses between words.

- Vocabulary
  Most speech recognition implementations rely on a particular set of speech grammar rules, which limits their vocabulary to a particular set of words. Implementations vary in the size of the vocabulary they can support and whether they are limited to a predetermined vocabulary and grammar.

The media service interface is used to specify, as necessary, the speaker and/or the vocabulary and grammar, and to deliver the text corresponding to the recognized speech.

Fax Printer

Fax printer media services refer to the fax receive-and-print functionality available in a fax machine. If a telephone system connects a fax printer media service to a call on which the presence of a fax CNG tone is indicating an attempt to transmit a fax, the fax will be received and printed on the appropriate device.

Fax Scanner

The *fax scanner* media service refers to the fax scan-and-send functionality available in a fax machine. It is the complement to the fax printer media service. If a telephone system connects a fax scanner media service, the fax scanner media service will attempt to establish a modulated fax data connection with another fax-capable device on the call, and then will transmit any sheets of paper fed into the fax machine's paper scanner.

Fax Modem

Fax modem media services provide fax data modulation for sending and receiving fax transmissions. (See the sidebar "Modulated Data" on page 79.) The media service interface is used to send and receive the compressed image data and fax transmission control information.

Data Modem

Data modem media services provide data modulation for establishing bidirectional modem communication. (See the sidebar "Modulated Data" on page 79.) The media service interface is used to configure the modem service and to send and receive asynchronous data.

Digital Data

A *digital data* media service provides access to the raw stream of digital data associated with a digital data media stream. The media service interface is used to convey the data.

Video Phone

A *video phone* media service is analogous to the fax scanner and fax printer media services, but applies to media streams containing video data. When attached to a media stream, this media service displays video on the video screen associated with the appropriate device, and captures and transmits video from a camera associated with the device.

3.8. Logical Device Elements and Appearances

Every device that is able to participate directly in a call has a portion referred to as a *logical element*. The concept of logical and physical device elements was illustrated in Figure 3-30.

A logical device element is a resource that is used to manage call control and switching functionality. Such elements represent the media stream channels and signal processing facilities used by a device while participating in calls. All telephony features and services, except those that specifically act on components of a physical device element, act on the logical element of a device or a set of logical elements in a device configuration. (Device configurations, or the relationship between multiple logical and physical elements, are described in section 3.9.)

 A logical device element has a set of one or more *call appearances* that represent a device in a given connection. The rectangular symbol that was introduced earlier to represent devices in graphical notation reflects the participation of that device through a particular appearance within the corresponding logical device element.

Logical device elements have the following attributes:

- Appearance type
- Appearance addressability

All of the appearances within a particular logical device element have the same type and addressability.

3.8.1. Call Appearances

A call appearance, or just *appearance* for short, may be thought of as a "connection handler" within a logical device element. One appearance can be associated with, at most, only a single call (through a connection) at a time.

All switching services that manipulate a device's association with a call act on the appropriate appearance within the logical element of that device. Appearances inherit all of the properties, attributes, and feature settings that apply to their logical device element.

The concept of appearances within logical device elements is illustrated in Figure 3-37. In this case logical device element D1 (the logical element portion of device D1) has two appearances labeled "A1" and "A2." Appearance A1 is representing device D1 in connection D1C1, and appearance A2 is representing device D1 in connection D1C2.

Figure 3-37. Logical device elements and appearances

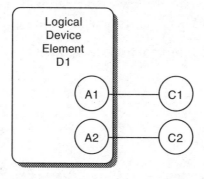

The state of an appearance is the state of the corresponding connection. If no connection is present at a given appearance, it is said to be *idle*.

3.8.2. Addressability

The *addressability* of an appearance refers to whether or not call processing is able to act specifically on a particular appearance within a logical device element.

Addressable Appearances

If an appearance is *addressable*, call processing can explicitly manipulate and observe connections associated with the appearance.

Addressable appearances are permanent and fixed in number. A logical device element that has addressable appearances cannot create new ones or destroy any. At any given time, a particular addressable appearance may or may not have an associated call.

Non–addressable Appearances

If an appearance is *non–addressable*, call processing is unable to distinguish a particular appearance from the encompassing logical device. In this case, the logical device itself is responsible for associating each connection with one of its appearances.

Non–addressable appearances are created and destroyed as needed to handle new connections that are associated with a particular logical device element. This behavior is illustrated in Figure 3-38.

Figure 3-38. Non–addressable appearance behavior

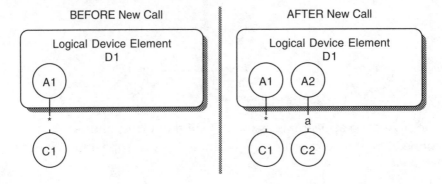

3.8.3. Appearance Types

The type of appearances supported by a particular logical device element plays a very significant role in determining the functionality and behavior associated with that device. There are two appearance types:

- Standard
- Bridged

The appearance type supported by a logical device element determines if it may be associated with, at most, one single physical device element, or if it may be shared by multiple physical device elements.

There are six behavior-type combinations (listed in increasing order of complexity):

- Selected–Standard
- Basic–Standard
- Exclusive–Bridged
- Basic–Bridged
- Independent–Shared–Bridged
- Interdependent–Shared–Bridged

Standard Appearances

Logical device elements containing *standard appearances* may not be shared by multiple physical device elements, and thus also are referred to as *private appearances*. All of the media streams associated with calls corresponding to the standard appearances are available only to the physical device element (if any) associated with the logical element in question.

Logical device elements with standard appearances exhibit one of two behaviors referred to as *selected* and *basic*.

Selected–Standard Appearance Behavior

When calls are presented to a logical device element that has *selected–standard* appearances, the new call is presented only to a single "selected" idle appearance.

Selected–standard behavior is illustrated in the example shown in Figure 3-39. In this example, D2 is a logical device element with four addressable standard appearances. Connection D2C1 is being handled by appearance A2. When a new call, C2, is presented to the device, selected–standard behavior dictates that a new *alerting* connection be created for a selected appearance, in this case A3.

Figure 3-39. Selected–standard appearance behavior

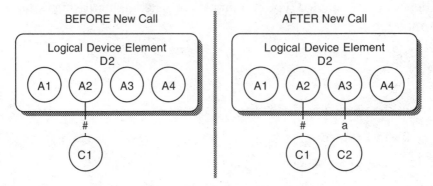

Basic–Standard Appearance Behavior

When calls are presented to a logical device element that has *basic–standard* appearances, the new call is presented simultaneously to all of the idle appearances.

When the call is answered by one of the appearances (i.e., the connection state of the corresponding connection transitions from *alerting* to *connected*), all of the other connections associated with the other standard appearances are cleared and the appearances return to being idle.

Basic–standard behavior is illustrated in the example shown in Figure 3-40. In this example, D3 is a logical device element with four addressable standard appearances. Connection D3C1 is non-*null* and is handled by appearance A3. When a new call, C2, is presented to the device, basic–standard behavior dictates that *alerting* connections are established for each of the (previously) idle appearances. After appearance A1 answers[3-27] the call, the connections that were created for appearances A2 and A4 are cleared.

Figure 3-40. Basic–standard appearance behavior

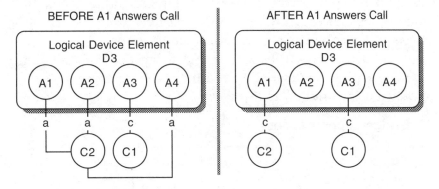

Bridged Appearances

Bridged appearances differ from standard appearances in that logical elements with bridged appearances may be associated with multiple physical device elements.

In the case of addressable bridged appearances, the logical device element can be associated with as many physical device elements as it has bridged appearances. Each bridged appearance is permanently associated with a particular physical device. The media stream associated with a call at a particular bridged appearance is available only to the corresponding physical device element.

In the case of non–addressable bridged appearances, the logical device element can be associated with any number of physical device elements, and it creates new appearances as needed. Each bridged appearance is associated with a particular physical device as long as a connection is present.

3-27. Answer switching service — The *answer* switching service will be described fully in Chapter 4. It involves taking a connection in the *alerting* or *queued* state and making it active by transitioning it to the *connected* state.

Typically each bridged appearance in a logical element is associated with a different physical device element, but this is not a requirement.

Bridged appearances are illustrated in Figure 3-41. In the example shown, logical device element D4 has three bridged appearances. It is associated with physical device elements D1, D2, and D3 through bridged appearances A1, A2, and A3, respectively. The association between a device and an appearance is shown graphically using double-headed arrows. The media stream associated with call C1 is accessible to physical device D1 by virtue of the fact that it is connected using bridged appearance A1.

Figure 3-41. Bridged appearances

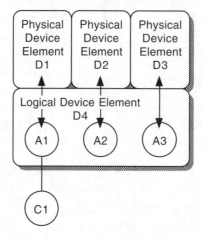

Logical device elements with bridged appearances exhibit one of four behaviors:

- Basic

- Exclusive

- Independent–shared

- Interdependent–shared

Basic–Bridged Appearance Behavior

When a new call is presented to a logical device element that has *basic–bridged* appearances, the call is simultaneously presented to all of the appearances.

When the call is answered by one of the appearances (the connection state of the corresponding connection transitions from *alerting* to *connected*) all of the other connections associated with the other bridged appearances are cleared.

Basic–bridged behavior is illustrated in the example shown in Figure 3-42. In this example, D4 is a logical device element with four bridged appearances. When a new call, C1, is presented to the device, basic–bridged behavior dictates that *alerting* connections are established for each of the (previously) idle appearances. After appearance A4 answers the call, the connections that were created for appearances A1, A2, and A3 are cleared.

Figure 3-42. Basic–bridged behavior

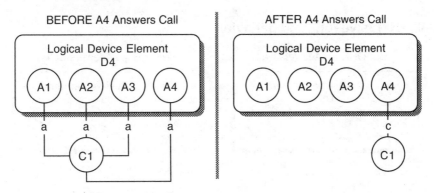

Exclusive–Bridged Appearance Behavior

When a new call is presented to a logical device element that has *exclusive–bridged* appearances, the call is presented simultaneously to all of the appearances.

When the call is answered by one of the appearances (i.e., the connection state of the corresponding connection transitions from *alerting* to *connected*), all of the other connections associated with the other exclusive–bridged appearances are *blocked* from further use until the connection to the appearance that answered is cleared.

Exclusive–bridged behavior is illustrated in the example shown in Figure 3-43. In this example, D4 is a logical device element with four bridged appearances. When a new call, C1, is presented to the device, exclusive–bridged behavior dictates that *alerting* connections are established for each of the bridged appearances. After appearance A4 answers the call, the connections that were created for appearances A1, A2, and A3 are blocked from further use by transitioning to the *fail* state.

Figure 3-43. Exclusive–bridged behavior

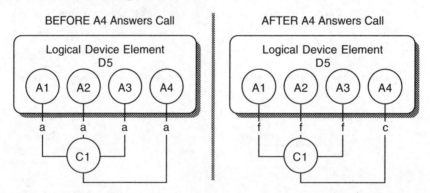

If the exclusive appearance places its connection to the call on *hold*, all of the connections at the bridged appearances transition to *hold*. Exclusivity is restored once again when any one of the connections transitions to the *connected* state and the others transition to the *fail* state.

Shared–Bridged Appearance Behavior

Shared–bridged appearances behavior (also known as *shared bridging*) has two cases, referred to as *independent* and *interdependent*. Shared bridging is the most important form of bridged appearances because it is the most common.

As with the other cases of bridging, when a new call is presented to a logical device element that has shared–bridged appearances, the call is presented simultaneously to all of the appearances.

In the case of shared bridging, however, when one appearance answers a call it becomes the appearance *participating*[3-28] in the call; the other shared–bridged appearances become *inactive*. The connections associated with the inactive appearances all transition to the *queued* state. This is illustrated in Figure 3-44. In this example, A4 answers the call and the connections handled by A1, A2, and A3 all transition to the *queued* state.

Figure 3-44. Shared–bridged behavior

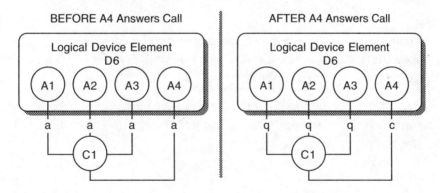

The key feature of shared bridging is that any of the appearances that are in the inactive mode (i.e., are handling a connection in the *queued* state) can answer at any time and be added to the call. Figure 3-45 continues the example from above. Shared–bridged appearance A2 answers call C1 and joins A4 as another participant in the call.

3-28. Participation — An appearance is considered to be *participating* in a call if its connection is in the *connected* or *hold* state.

Appearances A1 and A3 remain in the *queued* state. Additional appearances can be added to the call, up to a limit set by the telephone system implementation.

Figure 3-45. Shared–bridged behavior: adding a second appearance

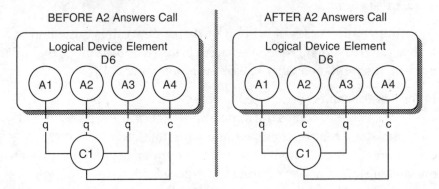

When an appearance drops[3-29] out of a call, it becomes *inactive* and the corresponding connection transitions to the *queued* state. This is illustrated in Figure 3-46. Here the example continues with A4 dropping from the call.

The call remains with the set of shared–bridged appearances represented by a given logical device element as long as at least one of the appearances is participating in the call.

The two forms of shared–bridged appearance behavior, *independent* and *interdependent*, differ in how appearances are affected when another appearance redirects a call or puts a call on *hold*.

3-29. Clear connection switching service — Dropping from a call is synonymous with clearing a connection to the call. The *clear connection* switching service will be described fully in Chapter 4. Normally the *clear connection* service transitions connections to the *null* state; in the case of shared–bridging, however, it transitions them to the *queued* state if other appearances in the same logical element are participating in the call.

Figure 3-46. Shared–bridged behavior: dropping an appearance

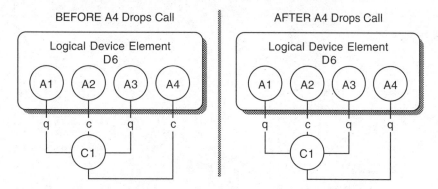

Independent–Shared–Bridged Appearance Behavior

In the case of *independent–shared–bridged* appearance behavior, when an appearance that is participating in a call moves the call away from the device, it has no effect on the other appearances unless it was the last participating appearance in the logical device element on the call. If it was the last, then the call is dropped from the whole logical device element. These two cases are illustrated in Figure 3-47 and Figure 3-48.

Figure 3-47. Independent–shared–bridged behavior: A2 and A4 active

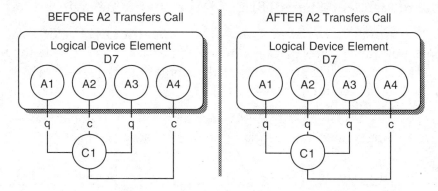

In Figure 3-47, both A2 and A4 are participating in the call when A2 transfers[3-30] it. Afterwards, A2 is *queued* and D7 remains associated with the call.

Figure 3-48. Independent–shared–bridged behavior: only A2 is active

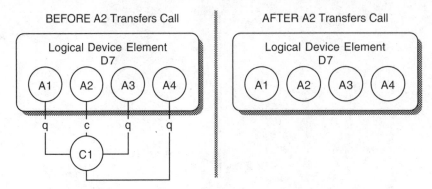

In Figure 3-48, only A2 is participating in the call when A2 transfers it. Afterwards, the call has left device D7 entirely.

Hold[3-31] works similarly in the case of independent–shared–bridged appearances. If an appearance places a call on *hold*, and there are other participating appearances in the logical device element on the call, then only the connection corresponding to the appearance in question is affected. All the connections are placed in the *hold* state only if the appearance was the last participating one.

Interdependent–Shared–Bridged Appearance Behavior

In the case of *interdependent–shared–bridged* appearance behavior, when any appearance that is participating in a call moves the call away from the device, the call is dropped from the whole logical device element. This is illustrated inFigure 3-49.

Hold works similarly. If any appearance places a connection in the *hold* state, all the connections are placed in the *hold* state.

3-30. Transfer switching service — The *transfer* switching service will be described fully in Chapter 4, section 4.12. It involves moving a call to a different device.

3-31. Hold switching service — The *hold* switching service will be described fully in Chapter 4, section 4.11.1. It involves suspending the media stream associated with a particular connection.

Figure 3-49. Interdependent–shared–bridged behavior

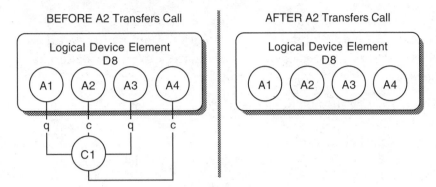

3.9. Device Configurations

A *device configuration* describes the arrangement of the various elements and appearances that can be associated with a given device. An endless variety of different device configurations may be formed from the possible combinations of physical elements, logical elements, and different appearance types.

Device configurations are described in terms of a specific *base* device. The following attributes determine the device configuration for a particular device:

- Physical element (Yes/No)
- Physical element's other associated logical devices
- Logical element (Yes/No)
- Logical element's other associated physical devices
- Logical element's addressability
- Logical element's behavior-type
- Maximum/total appearances

Collectively these attributes can be thought of as device configuration attribute for a specific device.

This section looks at a number of typical, or foundational, examples that illustrate how device configurations may be formed. (In Chapter 5 we will look at how these are applied in telephone system implementations.)

3.9.1. Logical Element Only

A *logical element only* device configuration consists, as the name suggests, of only a logical device element containing non–addressable standard appearances. Generally all devices other than station devices have *logical element only* device configurations.

The *logical element only* device configuration illustrated in Figure 3-50 has the following attributes:

- Physical element: No
- Physical element's other associated logical devices: N/A
- Logical element: Yes
- Logical element's other associated physical devices: None
- Logical element's addressability: Non–addressable
- Logical element's behavior-type: Selected–Standard
- Maximum/Total appearances: unlimited[3-32]

Figure 3-50. Logical element only device configuration

3-32. Maximum appearances — In a product implementation, the maximum number of non–addressable standard appearances is typically limited to some predetermined maximum.

144

3.9.2. Basic

A *basic* device configuration consists of a single physical device element associated with a single logical device element that contains non–addressable standard appearances. This device configuration applies to simple station devices.

The *basic* device configuration illustrated in Figure 3-51 has the following attributes:
- Physical element: Yes
- Physical element's other associated logical devices: None
- Logical element: Yes
- Logical element's other associated physical devices: None
- Logical element's addressability: Non–addressable
- Logical element's behavior-type: Selected–Standard
- Maximum/total appearances: Unlimited[3-32]

Figure 3-51. Basic device configuration

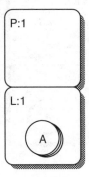

In this illustration, the labels "L" and "P" are used to denote logical and physical device elements respectively. The appearance "well" symbol represents a pool of non–addressable standard appearances.

Another variation of the *basic* device configuration involves two different devices, one with only a logical device element and one with only a physical device element, that are associated with each other. This is illustrated in Figure 3-52. From the perspective of the physical device element P1, the device configuration in this example can be represented as follows:

- Physical element: Yes
- Physical element's other associated logical devices:
    - L2 (Non–addressable/Selected–Standard)
- Logical element: No
- Logical element's other associated physical devices: None
- Logical element's addressability: N/A
- Logical element's behavior-type: N/A
- Maximum/total appearances: N/A

Figure 3-52. Basic device configuration consisting of two devices

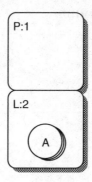

3.9.3. Multiple Logical Elements

A *multiple logical elements* device configuration consists of a single physical device element associated with multiple logical device elements that each contain standard appearances. This is a common device configuration for multiple line station devices.

One example of a *multiple logical elements* device configuration is illustrated in Figure 3-53. It has the following properties:
- Physical element: Yes
- Physical element's other associated logical devices:
    - L2 (Non–addressable/Selected–Standard)
- Logical element: Yes
- Logical element's other associated physical devices: None
- Logical element's addressability: Non–addressable
- Logical element's behavior-type: Selected–Standard
- Maximum/total appearances: 1

Figure 3-53. Multiple logical elements device configuration

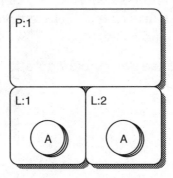

As we saw with the *basic* device configuration, none of the logical device elements in this device configuration need be part of the same device as the physical device element.

Multiple logical element device configurations allow a single physical element (a telephone set) in a telephone system supporting only one appearance per logical device element to have access to multiple calls simultaneously. Other telephone system features usually are set so that if a new call arrives for a logical element with an appearance that is not idle, it will be redirected to another logical device element within the same device configuration.

3.9.4. Multiple Appearance

A *multiple appearance* device configuration consists of a single physical device element and a single logical device element containing two or more addressable standard appearances.

One example of a *multiple appearance* device configuration is illustrated in Figure 3-54. It has the following attributes:
- Physical element: Yes
- Physical element's other associated logical devices: None
- Logical element: Yes
- Logical element's other associated physical devices: None
- Logical element's addressability: Addressable
- Logical element's behavior-type: Selected–Standard
- Maximum/total appearances: 3

Figure 3-54. Multiple appearance device configuration

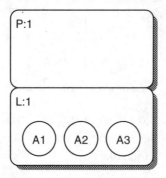

Multiple appearance device configurations are another way to allow a single telephone set to have access to multiple calls simultaneously.

3.9.5. Bridged

A *bridged* device configuration involves, as the name implies, bridged appearances. The form of a *bridged* device configuration therefore depends on whether the base of the device configuration is a physical or logical element.

In the example presented in Figure 3-55, the device configuration shown is for logical device element L3, which has bridged appearances. It has the following attributes:

- Physical element: No
- Physical element's other associated logical devices: N/A
- Logical element: Yes
- Logical element's other associated physical devices:
    - P1 (using appearance A1)
    - P2 (using appearance A2)
- Logical element's addressability: Addressable
- Logical element's behavior-type: Shared–Bridged
- Maximum/total appearances: 2

Figure 3-55. Bridged device configuration for a logic device element

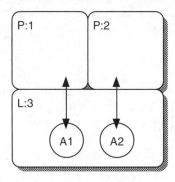

149

The device configuration for one of the physical device elements in this example is shown in Figure 3-56. It has the following attributes:

- Physical element: Yes
- Physical element's other associated logical devices:
    - L3 (using bridged appearance A1)
- Logical element: No
- Logical element's other associated physical devices: N/A
- Logical element's addressability: N/A
- Logical element's behavior-type: N/A
- Maximum/total appearances: N/A

Figure 3-56. Bridged device configuration: for a physical device element

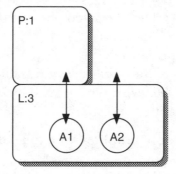

3.9.6. Hybrid

A physical device element associated with multiple logical device elements, each of which have different types of appearances, is said to have a *hybrid* device configuration.

In practice, most station devices are one of the typical device configurations described earlier. *Hybrid* device configurations are not at all uncommon, however, given the steady increase in telephone systems that support both standard and bridging appearance functionality simultaneously.

An arbitrary example of a *hybrid* device configuration is shown in Figure 3-57. It has the following attributes:

- Physical element: Yes
- Physical element's other associated logical devices:
    - L2 (with non–addressable standard appearances)
    - L3 (using bridged appearance A1)
    - L3 (using bridged appearance A2)
- Logical element: Yes
- Logical element's other associated physical devices: None
- Logical element's addressability: Addressable
- Logical element's behavior-type: Selected–Standard
- Maximum/total appearances: 3.

Figure 3-57. Hybrid device configuration

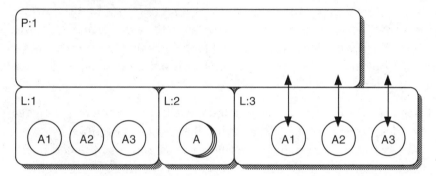

3.10. Addressing Devices

Now that we have seen how telephone systems, networks, and their resources allow the interconnection of device elements, and how complex device configurations can be formed, we turn to the subject of addressing. This refers to the means for referencing a particular device, element, or appearance of interest somewhere in a telephone network, or in a specific set of telephony resources.

3.10.1. Directory Numbers and Dial Plans

The addressing scheme for a telephone network is referred to as its *number plan* or *dial plan*. Telephone networks are hierarchical in nature, so each subset of a telephone network incorporates and expands on the dial plan of the larger network. Telephone networks that interconnect either encapsulate their neighbors' dial plans or provide a means of translating to them.

The address for a particular logical device in a telephone network is a *directory number* or, less precisely, a *phone number*.

If a telephone system allows a number of devices to share one or more network interface devices for access to an external network, then these devices are referred to as *extensions*. The extensions share the directory numbers assigned by the external telephone network to the shared network interface devices. Extensions have their own directory number inside the telephone system (typically one to five digits in length), but may or may not be directly addressable by devices outside the telephone system.[3-33]

A *prefix* is a digit or sequence of digits that indicates that a given *dial plan rule* is to be used for interpreting the rest of the sequence. Typically this is used to indicate that a call is to be established using a particular network interface device, carrier, or some other specific resource. For example, the digit "9" is often used to indicate that a sequence is to be interpreted as an external phone number.

3-33. DID and DIL — *DID* or *Direct Inward Dialing* is a service provided by the external network in which a whole block of directory numbers is associated with a particular network interface device (or group of network interface devices). Any call at that network interface device is then connected to a particular extension in the telephone system, based on the originally requested directory number. With *DIL*, or *Direct-In-Line*, all calls to a particular network interface device are always connected to the same internal device. For more information on DID and DIL, see Chapter 4, section 4.6.2 and Chapter 5, section 5.2.4.

3.10.2. Addressing in the Public Network

The International Telecommunications Union, or ITU, is responsible for setting standards for interoperability in the PSTN (the worldwide telephone network). The ITU has defined the basic elements that make up a telephone number. These are described below:

- Country code

  The *country code* portion of a telephone number is a one-, two-, or three-digit code that indicates the top-level network to which the number belongs. It refers either to a country like Bosnia (country code 387) or India (country code 91), for example, or a region like North America (country code 1) covered by an integrated numbering plan. The ITU is responsible for assigning[3-34] the unique country codes that make up the first tier of the worldwide numbering plan.

- Area code

  The *area code* portion (which might or might not apply to a given network, and is optional in certain networks) refers to a city or a particular region within the network specified by the country code. For example, the area code for British Columbia is 604 and the area code for Moscow is 095. The rules governing area codes are formulated uniquely within each network designated by a distinct country code. For example, in the North American numbering plan (NANP), area codes are always three digits long and, until recently, the middle digit had to be a "0" or a "1".[3-35] In some networks, area codes are all of uniform size; in others they may vary from area to area. In some networks every addressable device has an area code, but in others this not a requirement. In France, for example, until recently all numbers within Paris had area code 1 and all numbers outside of Paris had no area code.

3-34. Country code definitions -— Refer to ITU-T-E.163 for the official standard country code definitions. Your local telephone book should include an up-to-date list of country codes.

- Subscriber number

  The *subscriber number* specifies a particular logical device within a particular local network. The restrictions on the formation of subscriber numbers are once again specific to the network represented by a given combination of country code and area code. In the North American numbering plan, subscriber numbers are always seven digits.[3-36] A North American subscriber number might be 555-1234, but in a small village in China a subscriber number might be just 12.

- Sub-address

  The *sub-address* portion of a telephone number, which may or may not apply to a given network, refers to a particular device that is sharing a subscriber number with other devices through bridging. It allows reference to a specific bridged appearance.

Both ITU and ECMA (formerly the European Computer Manufacturing Association) also have defined a collection of digit sequence notations that define how telephone systems may interpret the sequences of digits representing telephone numbers exchanged between components in a telephone network. The specific notations are described in the sidebar "Standard Telephone Number Exchange Notations."

3-35. North American Numbering Plan — Until January 1, 1995, the number of area codes in the North American Numbering Plan was limited to 152 because 2 through 9 are the only legal values for the first digit (0 and 1 are reserved) and the second digit had to be 0 or 1. The restriction on the middle digit was applied to ensure that there would be no overlap between the set of numbers used as area codes and the first three digits of subscriber numbers (referred to as the central office code). After January 1, 1995, the "middle digit restriction" was relaxed, which resulted in a set of 792 area codes and expanded the overall potential size of the North American network from approximately 1 billion subscriber numbers to over 6 billion.

3-36. North American Numbering Plan — The first three digits of a subscriber number in the North American Numbering Plan are referred to as the *central office code* or *exchange code*. The next four digits are referred to as the *subscriber code*. All of the devices with a common exchange code are generally part of the same telephone system and have a common switching resource.

Standard Telephone Number Exchange Notations

ITU and ECMA specifications provide the following list of notations that telephony resources in a public or private telephone network can use to convey how addressing information associated with a call is to be interpreted. Each of these notations is associated with a raw, unpunctuated sequence of digits intended for machine, not human, consumption.

The notation names use the acronym *TON,* which stands for *Type of Number.* The formats defined are:

Implicit TON
Example: "01133112345678"

This notation indicates that the address is a digit string that includes all of the digits originally dialed to specify the destination for the call, including any prefixes. In this example the first three digits are the prefix for international direct dialing ("011"). The next two digits are the country code ("33" for France). This is followed by the single digit area code ("1" for Paris), and finally the eight digit subscriber number ("12345678").

Public TON – unknown

This notation indicates that the meaning of any digits presented is unknown.

Public TON – international number
Example: "33112345678"

This notation indicates that the digits presented include the complete digit sequence for an international telephone number. This is the same destination used in the first example, but in this case the sequence does not include prefixes. It is simply the concatenation of country code, area code, and subscriber number.

Public TON – national number
Example: "112345678"

This notation applies only to calls that remain within a country network, because it specifies that the country code is omitted. In this example, the country code is assumed, so the sequence is a concatenation of only the area code and subscriber number.

Standard Telephone Number Exchange Notations (Continued)

Public TON – subscriber number
Example: "12345678"

This notation is for local calls. In this example we are left with only the digits of the subscriber number; everything else is assumed.

Public TON – abbreviated
Example: "611"

This format is used when the public network supports an abbreviation of telephone numbers. This includes "super speed-dial" numbers and special network numbers. In this example a call originating in North America is being placed to a local carrier's repair number. By specifying that abbreviated notation is being used, the network recognizes "611" as a short form, or abbreviation, of the carrier's repair number.

Private TON – unknown; – level 3 regional; – level 2 regional; – level 1 regional; – local; – abbreviated
Example: "245645678"

There are private network versions of each of the public network formats described above. The only difference between the private and public versions are that the knowledge of the dial plan needed to decode the private network formats are specific to a particular private network or portion thereof. The digit string in the example might be used with any of these notations. Suppose the notation was specified as level 3 regional. The telephone system would need to know that for the private network in question the level 3 regional codes are 1-digit, the level 2 codes are 2-digit, the level 1 codes are 1-digit, and the remainder of the digits are a subscriber number. With this knowledge, plus the indication that level 3 regional notation was being used, the system would know that the number in question was: level 3 "2", level 2 "45", level 1 "6", and subscriber number "45678".

For more information on these standards, consult the following standards documents: ITU-T-E.160, ITU-T-E.131, and ECMA-155.

3.10.3. Dial Strings

When a call is initiated, the most common way to specify the desired destination device is through a *dial string* or a sequence of *dialable digits*. A sequence of dialable digits may include:

- Digits making up all or part of a destination device's directory number;

- Embedded commands that direct the telephone system to use a particular external network, carrier, or routing logic;

- Embedded commands that direct the telephone system to wait for a period of time to pass, wait for silence, wait for a dial tone, wait for billing tones, or wait for additional digits; and

- Digits that represent billing or other information distinct from the destination device address.

The characters that may be used in a dial string in a given instance depend on the interface being used for issuing the dial string and the subset of possible characters supported by a particular implementation. See the sidebar "Dialable Digits Format" for the complete set of characters in a dial string, their meanings, and some examples. The examples illustrate both the power of referencing devices using dial strings, as well as the disadvantages of this approach.

Dialable Digits Format

Dialable digits format consists of a sequence of up to 64 characters from the set of characters listed below.

Each character in the sequence represents a digit to dial or an embedded command of some sort. Every telephone system implementation supports the twelve characters found on all telephone keypads ('0' through '9', '*', '#') but the rest of the characters defined are optional.

The defined characters are:

'0'–'9' '*' '#' Digits to be dialed or DTMF tones to generate

'A'–'F' DTMF tones to generate

'!' Flash the hookswitch

'P' All digits that follow are to be pulse dialed

'T' All digits that follow are to be tone dialed

',' Pause

'W' Pause until dial tone is detected

'@' Pause until ringback followed by a period of silence is detected

'$' Pause until billing tone (bong tone) is detected

';' Pause for another dial string to follow

These characters can be placed into the sequence in any order or combination. For example, a sequence of multiple comma characters would result in a long pause.

The following are some typical examples of dialable digits format. In each case a call is being placed to the same destination: extension 789 behind the directory number 555-1234 in New York City.

Example: "789"

In this example the caller is using the same telephone system, so the call is internal to the telephone system itself.[a] The digit sequence consists of only the digits of the extension number "789".

Dialable Digits Format (Continued)

Example: "321789"

In this example, a call is being initiated from a telephone system in Portland that is part of the same private network as the telephone system in New York. The "321" indicates that a tie line[b] between the two telephone systems is to be used. This is followed by "789", the extension number. No embedded commands for pausing are needed because the private network allocates the tie line very quickly and all signaling is managed transparently.

Example: "19,12125551234@789"

In this example, the call is being placed from a public telephone in France. The digits "19" instruct the telephone system to allocate a network interface for an international call. The comma indicates that the dialing should be paused for the last embedded command to complete. The digit "1" indicates that the call is directed to North America, the digits "212" indicate the destination is New York City (specifically Manhattan), and the seven digits "555-1234" indicate the desired subscriber number. The "@" character tells the telephone system to wait until it detects that ringing has stopped, the call has been answered, and there is silence again. The telephone system at 555-1234 answers the incoming call using an auto-attendant that waits for the desired extension to be dialed. When silence is detected, the final three digits in the sequence, "789", are dialed.

Example: "9,1028802125551234$88844466669999@789"

In this example the caller is using a telephone system in Los Angeles. The digit "9" indicates that the telephone system should allocate the first available network interface device associated with the local carrier. As before, the comma means to wait; we need to give the previous command some time. The digits "102880" indicate that the call is to be placed using an AT&T credit card. The digits "2125551234" are the area code and subscriber number. The "$" indicates that the telephone system should pause until it hears the bong tone from the AT&T credit card system. The dialing then continues with the digits "88844466669999", which comprise the caller's credit card number. Finally the telephone system waits for the destination to answer and go silent; it then continues with "789", the extension number.

Dialable Digits Format (Continued)

Example: "4W5551234@789"

In this example, the call is being placed from a telephone system in Toronto. The "4" indicates that the telephone system is to use a special foreign exchange line that connects directly to the local network in New York City. Once again we need to pause while this is going on, however because the delays in setting up this connection vary, and because this system delivers dial tone once the connection to New York is established, the "W" command is used. The call is being placed as a local call on the New York City network (even though it actually is being initiated from Toronto), so the number dialed includes only the subscriber number "5551234". The example then proceeds as before, with a wait for silence and the dialing of the extension.

a. **Internal calls** — Internal calls are calls local within a private telephone system and are sometimes referred to as *intercom calls*.

b. **Tie lines** — A tie line is a dedicated transmission facility linking two telephone systems in a private or virtual private network.

A disadvantage of referencing devices using dial strings is that determining what dial string to use to refer to a particular destination depends on location, knowledge of the dial plan at that location, and knowledge of the supported subset of the dialable digits format implemented in the telephone system about to be used. A second disadvantage is that the embedded commands in dialable digits format are applicable to the placing of calls but not to monitoring or manipulating a device in some other way. This means that separate references to the same device may be expressed differently when using dialable digits format.

Dial strings are the only form of device reference that can be entered using any telephone keypad, however, so they are the one format, and in many cases the only format, supported by every telephone system.

3.10.4. Canonical Phone Numbers

Canonical phone number format was developed in order to overcome the limitations of dial strings. It allows for the specification of location-independent, absolute references to a particular device in a telephone network.

Canonical numbers are meant for use by people (on stationery, business cards, personal calendars, etc.), by computers (in scheduling programs, personal information managers, databases, etc.), and by telephone systems. As a result, the format uses punctuation that is easy for a person to write and for a computer or telephone system to interpret.

Phone numbers expressed in canonical representation can be absolute because they allow the inclusion of all the information necessary to identify a particular device, regardless of the context. For example, a particular device could be called from within the same telephone system, from a different telephone system, or from anywhere in the public network using the same canonical number. In each case, the information necessary to establish a call from the location in question is intelligently extracted from the canonical phone number.

A device reference in canonical phone numbers may include:

- A country code, area code, and subscriber number;

- A sub-address and/or extension number; and

- A name

Various portions of the canonical format are optional. Incomplete canonical numbers are interpreted based on other context information. For example, if a given canonical number included only a name, a telephone system with access to a directory could simply look up the name and establish a call to the corresponding person. Other portions not provided are appropriately interpreted as either not being applicable to the device, or as being the same for the location of the caller. See the sidebar "Canonical Phone Number Format" for the complete set of symbols used in a canonical phone number and their meanings.

Canonical Phone Number Format

Canonical phone number format consists of a sequence of up to 64 characters using the following syntax:

O< [**DPR1**, **DPR2**, ... , **DPRn**] +**CC** (**AC**) **SN** *__SA__ x**EXT**>**NM**

The bold items are symbols that are replaced by the appropriate data. The other symbols are delimiters used to punctuate the string. The different portions of the canonical number are as follows:

'O' The leading 'O' indicates that this number is in canonical format. In practice, the 'O' is only used when communicating canonical numbers electronically. People do not require this character, so it will not appear in what they see and use.

'<' '>' The angle brackets indicate that a name (**NM**) is included in the canonical number. If no name is present, the angle brackets are not present. If a name is present, they enclose the rest of the canonical number.

'[' ']' The square brackets indicate that a list of one or more *dial plan rule* strings (**DPR**s) is included in the canonical number. A dial plan rule specifies how the other portions of the canonical number are to be interpreted. Most often the canonical number refers to a device in the public telephone network, so no dial plan rule is required. If the device is in a private network, however, the dial plan rule identifies the private network involved. If multiple dial plan rules are required, they are separated by commas.

'+' The '+' character indicates the presence of a country code (**CC**).

'(' ')' The parentheses enclose the area code (**AC**) portion of the canonical phone number. The presence of a dial plan rule for a private network might indicate that this portion is to be interpreted as a subnetwork.

SN This represents the actual subscriber number portion of the canonical phone number. It is required for use with the public network.

'*' The '*' character indicates the presence of a sub-address (**SA**).

'x' The 'x' character indicates the presence of an extension number (**EXT**).

NM This represents the name associated with the device being referenced.

Canonical Phone Number Format (Continued)

All portions of a canonical phone number are optional. However, a canonical number is required to include at least a subscriber number, extension number, or name. If the area code is not included and there are no dial plan rules indicating this is a private network reference, the area code will be interpreted either as not applicable, or as being the same as that of the device interpreting the number (depending on the country code). If the country code is not included and there are no dial plan rules indicating this is a private network reference, the country code will be interpreted as being the same as that of the device interpreting the number.

The following are two examples of complete device references in canonical phone number format for extension 789 behind the directory number 555-1234 in New York City.

Example: "O+1(212)555-1234x789"

Example: "O<+1(212)555-1234x789>John Smith"

The following are other examples of device references in canonical phone number format:

Example: "O[MYNET](4)x40220"

In this example, a dial plan rule ("MYNET") indicates that the device referenced is in the private network MYNET. The area code ("4") is then interpreted as a particular subnetwork in MYNET and "40220" is a particular extension on that subnet.

Example: "O+33(1)12.34.56.78*06"

In this example, the canonical phone number refers to a particular bridged device with a sub-address of "06" at the subscriber number "12.34.56.78" in Paris (country code "33" and area code "1").

Example: "O555-1234"

In this example the canonical phone number is only partially specified. It is a valid device identifier in canonical format, but it requires the interpreter to assume the missing information. In this case, if the interpreter were in New York City, the country code would be assumed as "1" and the area code would be assumed as "212".

Example: "Ox40220"

This is another example of a partially specified device in canonical number format. In this case the device in question is extension "40220" in the telephone system being used.

Example: "O<>John Smith"

Canonical numbers support *call by name*, that is, specifying the destination as a name that is then resolved by a computer or by the telephone system.

At the moment, canonical representation is used primarily by computers, which translate them into appropriate dial strings for telephone systems. This will change as telephone systems become more intelligent.

3.10.5. Switching Domain Representation

Another phone number format is the *switching domain*[3-37] *representation*. This representation is used to specify a particular device, element, or appearance within a particular telephone system.

When this representation is used, the directory number notation (see the sidebar "Standard Telephone Number Exchange Notations") must also be specified along with the phone number to allow interpretation of the directory number portion.

This format for referencing devices is the preferred format for references to devices within a given telephone system (as opposed to those in an external network) because it allows precise references down to the level of specific appearances and agents. See the sidebar "Switching Domain Representation Format" for the complete set of symbols used in this format and their meanings.

3-37. Switching domain — The term *switching domain* is defined in Chapter 6.

Switching Domain Representation Format

Switching domain representation format consists of a sequence of up to 64 characters using the following syntax:

N<**DN** ***SA** &**CA** %**AID**>**NM**

The bold items are symbols that are replaced by the appropriate data. The other symbols are delimiters used to punctutate the string. The different portions of the switching domain representation are as follows:

'N' The leading 'N' indicates that this reference is in switching domain format.

'<' '>' The angle brackets indicate that a name (**NM**) is included in the device reference. If no name is present, the angle brackets are not present. If a name is present, they enclose the rest of the device reference.

DN This represents the directory number portion of the phone number.

'*' The '*' character indicates the presence of a sub-address (**SA**).

'&' The '&' character indicates the presence of a call appearance reference (**CA**).

'%' The '%' character indicates the presence of an agent ID reference (**AID**).

NM This represents the name associated with the device being referenced.

All portions of this representation are optional. However, a device reference in this format is required to include at least a directory number or agent ID.

The following are examples of device references in switching domain representation for extension 789 behind the directory number 555-1234 in New York City.

Example: "N789"

In this example, the given directory number notation is "Private TON - local." The digits "789" reference the device as an extension with directory number 789 in the telephone system in question.

Switching Domain Representation Format (Continued)

Example: "N789&02"

In this example, the given directory number notation is "Private TON – local." "02" references a specific call appearance on the device in the telephone system in question with the directory number identified by "789".

Example: "N<12125551234>John Smith"

In this example, the given directory number notation is "Public TON – international." This is the device reference that a telephone system outside of North America would use if a call from this device had been received. Extension 789 shares the network interface device identified in the external network by "5551234" so it would be referenced using that number across an external network.

Example: "N5551234"

In this example, the given directory number notation is "Public TON – subscriber." This is the device reference that another telephone system in New York City would use if a call from this device had been received. As with the previous example, "5551234" refers to the subscriber number used because extension 789 shares the network interface device and only this number is visible in the public network.

The following are other examples illustrating the use of this format in reference to devices within a particular telephone system. The given directory number notation is therefore "Private TON – local" in each case.

Example: "N%604"

This example is a reference to the device associated with agent ID "604".

Example: "<N%604>John Smith"

This example is a reference to the device associated with agent ID "604" and the name "John Smith".

Example: "N40220*02"

This example is a reference to bridged appearance "02" on the extension with number "40220".

3.10.6. Device Numbers

The fourth type of device reference is the *device number*. In most telephone systems, every single device is internally assigned a unique device number. Some implementations allow devices to be referenced directly using these numbers.

See the sidebar "Device Number Format" for a description of the components of this format.

Device Number Format

Switching domain representation format consists of a sequence of up to 64 characters using the following syntax:

\N

The bold items are symbols that are replaced by the appropriate data. The other symbols are delimiters used to punctuate the string. The different portions of the switching domain representation are as follows:

'\' The leading '\' indicates that this reference is in device number format.

N This represents the number associated with the device being referenced.

All portions of this representation are optional. However, a device reference in this format is required to include at least a directory number or agent ID.

3.11. Review

In this chapter we have seen that a *telephone system* may range in scope from an individual telephone to a vast telephone network. Telephone systems represent a collection of *telephony resources* that also may be referred to as a *telephony resource set*. Types of telephony resources include *call processing, switching, interfaces, devices*, and *dynamic entities*.

Calls are dynamic entities in a telephone system that represent a media stream and associated control information traveling between two or more points.

Devices are resources responsible for consuming and generating the media streams and control information associated with calls. A *network interface device* can be used to establish calls between one telephone system and a larger telephone network. Other device *types* include *station, park, pick group, ACD, ACD group*, or *hunt group*. Devices may include a *logical element*, or a *physical element*, or both.

Physical device elements consist of *components* that make up the physical interface of a device. These components include *hookswitches, auditory apparatuses, buttons, lamps, message waiting indicators, displays*, and *ringers*.

Logical device elements represent sets of *call appearances* of a particular type. Each call appearance (*appearance* for short) that a logical element possesses enables it to participate in a call. Appearances have associated *types* and *behaviors*. The set of type-behavior combinations are *dynamic, basic–static, selected–static, basic–bridged, exclusive–bridged, independent–shared–bridged*, and *interdependent–shared–bridged*. Appearances may be *addressable* or *non–addressable*. An appearance that is not involved in a call is *idle*. Exclusive–bridged appearances may be *blocked* from a call and shared–bridged appearances may be either *participating* or *inactive* with respect to a call.

Device configurations describe the relationship between a particular device element and the device elements associated with it.

Device addressing refers to the ways that a particular device or appearance can be referenced. Addressing formats include *dial strings*, *canonical phone numbers*, *switching domain representation*, and *device numbers*.

Connections define the relationship between a particular call and a particular call appearance. The most important attribute of a connection is its *connection state*, which may be *null*, *initiated*, *alerting*, *connected*, *hold*, *queued*, or *fail*.

Media stream channels are allocated for connections as needed and are responsible for conveying the associated media streams.

Switching resources are responsible for carrying out services that create, clear, and manipulate the states of connections, and for allocating and deallocating media stream channels as needed. These *switching services* can be illustrated through a simple graphical notation that shows the devices, calls, connections, and connection state transitions involved.

Call processing is primarily responsible for *routing* or directing calls through the telephone system by managing the switching resources. It takes action based on feature settings, timers, default rules, and commands received from devices and through telephone system interfaces. Certain types of devices exist within a telephone system in order to provide specialized routing functionality.

Now that a complete abstraction of telephony resources and their attributes has been defined, we can explore the features and services that can be implemented by manipulating these resources in well-defined ways.

4.
Core Telephony
Features and Services

This chapter presents the core telephony features and services commonly available in telephone systems independent of any CTI interface. These are referred to as *call control*, *call associated*, and *logical device* features and services. (Telephony features and services that are specific to a CTI interface are covered in Chapter 6.)

The preceding chapter explored a universal abstraction of telephone systems as telephony resource sets. Now we will breathe some life into these resources by seeing how they work together to deliver standard telephony features and services. As was true of the telephony resources themselves, these features and services are the result of international and industry standardization efforts. The resulting abstraction presented here represents a superset of the most frequently used telephony functionality. In other words, every vendor's product has a different combination of features that differentiates their product, but all of the significant features are represented here. Every telephony vendor has at least a few unique names for standard features, so popular alternative or proprietary terms are also referenced wherever appropriate.

Telephony features and services available to a given device or telephone system depend upon the telephony features and services provided by the telephone system or by an external network respectively. Making telephony features available (and billing for them if appropriate) is referred to as *subscribing* to a feature or service. The set of features and services available to a given device from its telephone system, or to a telephone system from an external network, is generally referred to as its *class of service*.

4.1. Basic, Supplementary, and Extended Services

Telephony services are operations that a telephone system may perform in response to commands from the system's devices or from a CTI interface. (Additional services that are specific to the CTI interface are presented in Chapter 6.)

Services are frequently grouped into three classes:

- Basic

  Basic telephony services refer to the lowest common denominator for all telephone service, specifically the ability to place, answer, and hang up calls.

- Supplementary

  Supplementary services refer to all the other well-defined services beyond the basic services.[4-1]

- Vendor specific extensions or Extended

  Vendor specific extensions, also called *extended* services, are the services beyond the supplementary services unique to a particular vendor or product.

4-1. Service naming — The term *supplementary services* specifically applies to the functionality of the services. The names used by different telephony and operating system vendors to refer to specific supplementary services vary. This book uses the terminology established by ECMA, Versit, and other standards-setting groups. Wherever appropriate, alternate names are provided.

This chapter presents the basic services, followed by the most popular supplementary services. Features unique to the products of a particular vendor are not represented here (though CTI interface support for vendor specific extensions is described in Chapter 6.)

4.2. Features

In addition to a telephone system's ability to respond to commands for switching services, the telephone system may track certain pieces of information about a call or device, manage timers associated with certain activities, or establish settings that govern call progress in some way. These are all referred to as *features*.

The term *feature interaction* refers to the ways that normal call progress may be affected by the certain features that may be in effect for a particular device.

In this chapter we will be describing all of the most popular telephony features in terms of the concepts presented in Chapter 3.

4.3. Basic Services

Basic services often are described as being the telephony functions available with a POTS[4-2] phone. It is the lowest common denominator for all telephone service:

- Placing calls

- Answering calls

- Hanging up calls

This section will explore each of these basic services. Later we will take a closer look at each one with respect to its use in the context of richer supplementary services.

4-2. POTS — *POTS* stands for *Plain Old Telephone Service*. It refers to traditional tip-and-ring analog telephone lines with only dial-answer-hangup functionality. POTS service is described further in Chapter 5.

4.3.1. Make Call

The formal name for the service that places calls is *make call*. This service, in its simplest form, does the following:

- Creates a new call

- Creates a new connection between the *calling device* and the call

- Routes the call to the destination (the *called device*) specified

In the basic case of the *make call* service, the operation is complete when the calling device has a connection to the new call in the *connected* state[4-3] and the called device has a corresponding connection in the *alerting, connected, fail, null,* or *queued* state. This is referred to as the service's *completion criteria*. Every switching service has a set of completion criteria that can be used to determine if the service succeeded or not.

A *make call* service may not succeed (or complete) for a variety of reasons. The most likely is that the calling device is already in use with another call and is therefore unable to place a new call.

 It is very important not to confuse the concept of a telephony service that does not succeed with the concept of the *fail* connection state. The connection state of *fail* indicates that call progress associated with a particular call has stalled and that a certain appearance may be blocked from the call. This is quite distinct from a telephone system's switching resources being unable to perform a specific operation.

After the *make call* has completed, the telephone system will attempt to connect the new call to the specified destination. Due to feature interactions and the effects of other routing mechanisms, however, the call could be redirected many times before actually arriving at a device

4-3. Make Call basic case — The completion criteria for the basic case of the *make call* service is that it transitions to the *connected* state. In the general case, however, it may transition to (or retransition to) the *initiated* state in order to support multi-stage dialing. This is explained in section 4.4.2.

where it can be answered. The call also could get stalled, for example, because the telephone network is congested or because the destination device is busy.

Make call is described in graphical notation in Figure 4-1.

Figure 4-1. Make Call service (single-step dialing)

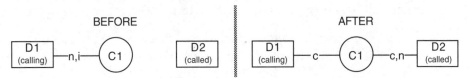

In the simplest case of the *make call* service, there is initially no connection (i.e., the connection is implicitly in the *null* state) between the calling device (D1) and the new call C1. (In fact, the new call doesn't yet exist.) There is a second, more common case, however, where the call already exists and the connection is in the *initiated* state. The difference between the two cases is that the first represents *on-hook dialing* and the second represents *off-hook dialing*. An example of on-hook dialing is what you are doing when you place a call from a cellular phone: You dial the number and press the Send button. Off-hook dialing happens when you pick up the receiver (or go off-hook in some way), hear dial tone, and then begin dialing. With off-hook dialing, the new call is created when you hear dial tone. The dial tone indicates that the connection is in the *initiated* state.

The *make call* service is complete in either case when the connection to the calling device transitions to the *connected* state as shown. At this point the device is said to have *originated* the call. The connection to the called device (D2) may be in any of the following states:

- *Alerting* – the call has already reached D2 and the telephone system is attempting to connect the call.

- *Connected* – the call not only has reached the called device, but has already been answered, possibly because D2 has the auto-answer feature[4-4] active. (An alternative case is when the call is an external call and it is connected to a network interface device as a proxy for D2. Making external calls is explained in section 4.4.3.)

- *Fail* – call processing stalled trying to connect to D2, most likely because it was busy or no network interface device was available.

- *Queued* – the call is being queued at the called device, probably because it is some type of distribution device.

- *Null* – the call was redirected so there is no connection to D2. There probably is a connection to the new call at some other device, however.

4.3.2. Answer Call

The service that answers calls is appropriately named *answer call*. This service is quite simple: It causes a connection in the *alerting* (or *queued*) state to transition to the *connected* state. The service also activates any appropriate auditory apparatus.

Typically the *answer call* service is triggered by picking up a telephone handset or pressing an appropriate hookswitch button to activate a headset or speaker phone auditory apparatus. This service also can be supported through the CTI interface if the physical element involved has a hookswitch that can be set to off-hook automatically by the telephone system.

The *answer call* service may not succeed for a variety of reasons. Two likely cases could be that the connection is in the wrong initial state (not *alerting* or *queued*), or that the answering device does not support going off-hook automatically.

Answer call is presented graphically in Figure 4-2.

4-4. Auto-answer — The *auto-answer* feature is described in section 4.9.1.

Figure 4-2. Answer Call service

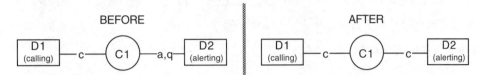

Before the service connection D2C1, the answering connection was in the *alerting* or *queued* state. After the call is answered it is in the *connected* state.

4.3.3. Clear Connection

Dropping a call, or "hanging up" on a call, is accomplished by clearing the associated connection. The service for doing this is appropriately named *clear connection*. This service removes just one connection from a multi-way call and is also used to inactivate a shared–bridged appearance.

The *clear connection* service transitions a connection from any non-*null* state to the *null, queued* state and sets the hookswitch on the physical device element to on-hook if appropriate. When the service completes, ordinarily the connection will be cleared and will transition to the *null* state. The exception is when the connection is one of a set being used by shared–bridged appearances. In this case, the connection will be cleared only if it corresponds to the last participating appearance in the set; otherwise it will transition to the *queued* state. (This behavior was described in Chapter 3, section 3.8.3.)

After the service completes, the other connections in the call are unaffected, assuming there are two or more connections remaining. If only one connection remains in the call after the service completes, the call itself may be cleared, in which case the remaining connection also would transition to the *null* state.

The *clear connection* service is illustrated in Figure 4-3. Before the service the *clearing* connection is D3C1, which is a connection in any non-*null* state. The other connections in the call may be in any state.

After the service completes, the D3C1 is in the *null*, *queued*, or *fail* state and the other connections are unaffected. (The *clear connection* service is discussed in further detail in section 4.15.1.)

Figure 4-3. Clear Connection service

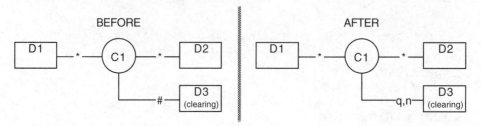

4.4. Placing Calls

Now that we've looked at all the basic services, we'll explore further the additional features associated with creating new calls. This section deals with the general case of the *make call* service (including prompting and multi-stage dialing), the concepts surrounding *external calls*, the *make predictive call* service, and the *last number dialed* service.

4.4.1. Make Call and the Initiated State

The *initiated* connection state plays a number of important roles in the implementation of the *make call* service.

If the calling connection (D1) is in the *initiated* state, the call has already been created, so there is dial tone and no instructions have been provided yet for the call. This is referred to as *off-hook dialing*.

Under certain circumstances the *make call* service may invoke *prompting* for the calling connection, which is indicated by having the connection state transition, or retransition, to the *initiated* state. (Prompting is explained in section 4.4.6.)

Finally, if the dial string supplied to the *make call* service was incomplete, the service will initiate multi-stage dialing. In this case the service completes when the connection state of D1C1 transitions (or, retransitions) to the *initiated* state.

4.4.2. Dial Digits for Multi-Stage Dialing

Multi-stage dialing, or *delayed dialing*, refers to the situation where a dial string is provided in increments rather than as a single string of digits.

In section 4.3.1 we saw the case of the *make call* service for single-stage dialing. The other case of the *make call* service is shown in Figure 4-4. The difference between the version in Figure 4-1 and the one in Figure 4-4 is that in the multi-stage case of the service, connection D1C1 is in the *initiated* state at completion.

Figure 4-4. Make Call service

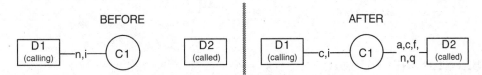

There are two ways to initiate multi-stage dialing.

1. Provide an incomplete dialing sequence to a *make call* service.

2. Manually take a telephone off-hook, which creates a new call with a connection in the *initiated* state.

In either case, additional digits in the dial string sequence are supplied using the *dial digits* service. The *dial digits* service is illustrated in Figure 4-5.

Figure 4-5. Dial Digits service

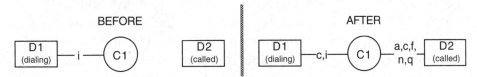

179

Connection D1C1 remains in the *initiated* state until a complete dial string has been provided. It then transitions to the *connected* state and the call is said to have been *originated*. Figure 4-6 shows a multi-stage dialing sequence.

Figure 4-6. Multi-stage dialing sequence

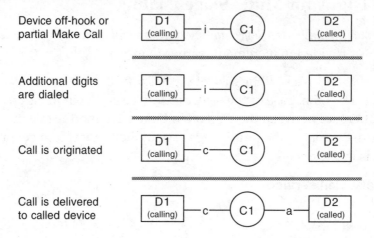

4.4.3. External Outgoing Calls

External calling is a feature that allows devices and calls inside a given telephone system to be connected to devices in an external network. (For a telephone system consisting of a single telephone, all calls are considered as external.)

Support for *external outgoing* calls is a feature that allows network interface devices to be connected to new calls being generated inside the telephone system. Network interface devices are the proxies within the telephone system that represent a distant called device's participation in a call. This is shown in Figure 4-7. In this illustration, the "?" symbol indicates that because the connection between the network interface device (D3) and the called device (D2) is entirely outside of the telephone system (and potentially is made up of many different connections), it cannot be controlled directly and the external network may or may not provide state information.

Figure 4-7. External outgoing call

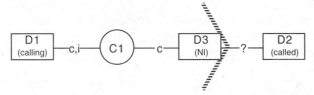

When the telephone system determines that a destination is external,[4-5] it routes the call through a network interface device. The point when it actually connects a network interface device to the call,[4-6] referred to as the *network reached* point, is shown in Figure 4-8.

Figure 4-8. Network reached

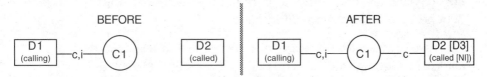

When dialing takes place in a single step using just the *make call* service, the call has already been originated, so connection D1C1 is in the *connected* state before the network interface device is connected. During multi-stage dialing, however, before receiving the entire dial string the telephone system may have enough of it to know that the call is an external outgoing call and place D1C1 into the *connected* state. In this case, D1C1 is still in the *initiated* state even though D2C1 is in the *connected* state.

4-5. Identifying external calls — A call may be identified as being for an external destination in a variety of ways. In some cases all calls are external, so the call is identified as being external automatically as soon as it is determined that a new call is being placed. Otherwise, the presence of a special prefix in a dial string or the information in a canonical phone number determines whether a particular destination is external or internal. See section 4.4.5 for more on dial plan management.

4-6. Seizing — When a network interface device is allocated for a call using the external network, it is known as *seizing the network interface* or *seizing the trunk*.

4.4.4. Network Interface Groups

If the telephone system has more than one station device, it typically has fewer network interface devices than stations. The stations share the network interfaces for more efficient utilization of these resources.

When an external outgoing call is made by a device that shares a pool of network interface devices, that call must be routed through the next available network interface. As we have already seen, a special device exists for just this purpose: The hunt group device redirects calls to the next available device in its group. A *network interface group*[4-7] is therefore a hunt group device that is set to distribute outgoing calls to the next available network interface device. If all network interfaces are busy, the call may either fail or wait for one to become available, depending on the implementation.

Figure 4-9 illustrates this sequence. In this example, D5 is placing an external outgoing call, C2. The new call is directed to the network interface group device D1. There are three network interfaces in the group: D2, D3, and D4. D2 is busy with call C1, so D1 finds the next available network interface device, D3, and redirects the call there.

4.4.5. Dial Plan Management and Least Cost Routing

The appropriate network interface device or group to use for placing a call, and the selection of an appropriate long-distance carrier (if appropriate), is determined from the address provided to identify the called device. The rules set up to manage the process of transforming a calling device's address into routing decisions for a call, and the process of performing these transformations, is referred to as *dial plan management*. One aspect of dial plan management is *least cost routing* functionality. Least cost routing involves determining the cheapest available way to place a given call, given the destination of the call, the

---

4-7. Trunk groups — Network interface devices are commonly referred to as *trunks*, so network interface group devices are commonly referred to as *trunk groups*.

Figure 4-9. Network interface device group behavior

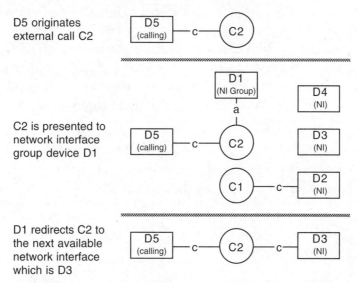

D5 originates
external call C2

C2 is presented to
network interface
group device D1

D1 redirects C2 to
the next available
network interface
which is D3

available private network and public carrier options for routing the
call, the location from which the call is originating, the date and time
of day at that location, and current rate plan information.

One or more of the following techniques typically are used to
implement dial plan management:

- Prefixes

- Leading-digit translation

- Canonical number translation

Depending on the techniques implemented for a given telephone
system, the users of the system may have to make more of the routing
decisions themselves.

Prefixes

Prefixes are the most commonly used mechanism for indicating that a
call is to a device on an external network. Prefixes vary from single to
multiple digits, and are used at the beginning of a dial string to
indicate that a particular network interface, network interface group,

or routing mechanism is to be used to route a particular call. In North America the most common prefix in private telephone systems is "9". This digit usually indicates that a particular call is an external outgoing call using the local carrier. Other prefixes may be set up to correspond to network interfaces associated with connections to private networks, specific carriers, or frequently called destinations. (Options for network interfaces are covered in more detail in Chapter 5.)

If a telephone system relies entirely on prefixes to perform dial plan management, most of the routing effort is left to the people using the telephone system. To place a call they must enter (in varying orders):

- The most appropriate prefix for a call. Often they just dial the prefix for the local carrier even though this may not be the most cost-effective routing.

- The appropriate carrier selection code, if appropriate. If they don't do so, and the call is a toll call, it will be routed by the default carrier.

- Any applicable billing codes

- The desired destination number in an appropriate form for the external network associated with the selected network interface device. Depending on the network interface used, the caller might not have to dial a long-distance prefix or an area code, or even the full subscriber number.

The scenario described above is the worst case, but is the scenario encountered by most users of hotel, small business, and home telephone systems.

Leading Digit Translation

Leading digit translation involves analyzing the first few digits of the called device's number.

Simple systems rely on just the area code to determine how to route a call. In North America this is referred to as *three-digit translation* because the first three digits of a ten-digit telephone number are its area code. These systems route all local calls to the local carrier and use least cost routing tables to determine the least cost route for each long-distance area code.

Many area codes in the North American dial plan have unique dialing rules for long-distance dialing within an area code and to adjacent area codes, however. Some require that the digit "1" be dialed and others do not. Some allow the number to be dialed with or without the leading "1" but will charge more for the call depending on how it is dialed. Some allow calls to be placed to different areas codes using seven-digit numbers. All these exceptions make it necessary for robust systems to be based on *six-digit translation* that relies on a database of dial plan rules.

Dial plan rules vary dramatically throughout the world, and leading-digit translation typically must be designed for each country individually, so a system built for use in one country will not work in another.

Canonical Number Dialing

Canonical number dialing, or full-number dialing, means working with numbers in canonical form (as described in Chapter 3, section 3.10.4) rather than just scanning leading digits. Dial plan managers built on canonical numbers may support the parsing of arbitrary numbers into canonical numbers based on context rules, but then do their routing and translation based on the resulting canonical number.

By working from canonical numbers, network interface devices can be added and removed with ease, and a single international dial plan database can be constructed and maintained more easily. Most important, users of the system always can dial a canonical number rather than worry about the right prefixes to use.

Using Dial Plan Management

Dial plan management and least cost routing are a significant area where CTI solutions can play a significant role, both in making telephone systems of all sizes easier to use and in reducing telephone expenses. Examples include:

- An individual traveling with a notebook computer can use CTI functionality to dial calls from hotel rooms, airports, and meeting locations. Using the dial plan management capabilities of the CTI software in the personal computer—rather than having the traveler figure out access codes, carrier codes, billing codes, etc.— saves a great deal of time and money.

- A home business owner can use CTI software in a personal computer to automate all outbound dialing. Although the home might have only two phone lines, the CTI software ensures that every voice, fax, and data call that is made is dialed correctly, uses the cheapest carrier given the time of day and the destination, and is placed on the most appropriate line (business versus personal).

- Larger telephone systems typically have dial plan management built in. Unfortunately, as the number of new area codes, country codes, carriers, and routing options increases, these built-in dial plan management features must be reprogrammed frequently. Using a CTI-based dial plan management system greatly simplifies the task of updating the dial plan information and provides much greater control to the telephone system's manager.

4.4.6. Prompting

Prompting refers to a feature that allows the *make call* service and certain other services to be used for on-hook dialing on station devices that don't have the ability to go off-hook without manual intervention. Prompting will delay the point where the call is originated until the device goes off-hook. Figure 4-10 shows the sequence of transitions for prompting in the case of the *make call* service.

Figure 4-10. Prompting feature

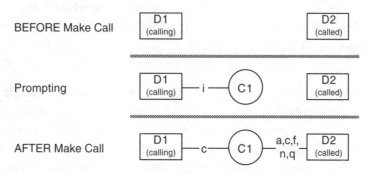

In this sequence, initially there is no connection (and no new call) because the *make call* is being dialed on-hook. When the dialing is complete, the call is created and connection D1C1 is placed in the *initiated* state while prompting takes place. When the station device is then taken off-hook manually, connection D1C1 transitions to the *connected* state, signifying that the call has been originated and satisfying the completion criteria of the *make call* service.

Prompting involves a telephone set making an indication, typically audible, to indicate that someone should take it off-hook in order to progress out of the *initiated* state. It is distinct from ringing, which is an indication made by a device that a connection is in the *alerting* state. Prompting is used whenever some service is initiated for a device but it is unable to go off-hook automatically.

4.4.7. Make Predictive Call

The *make predictive call* service is a very powerful capability that allows the telephone system to make calls on behalf of the calling device without actually involving it in the call until the call is delivered or connected to the desired destination and other criteria are satisfied.[4-8]

4-8. Predictive dialing — The name of the *make predictive call* service derives from the fact that the device or computer requesting the service must predict the availability of someone to handle the call if and when it completes, based on call completion and call duration statistics.

This service typically is used in situations where one or more people are making large numbers of calls and want to spend every possible moment talking to "live" people, and not be tied up dialing telephones, waiting for the calls to be placed, waiting for the calls to be answered, getting answering machine recordings, etc.

With predictive dialing, the telephone system typically is instructed to make the calls automatically and to wait until the call is either delivered to, or answered by, the called device. If the option to wait for the call to be *connected* is chosen, the telephone system may be further instructed to differentiate between a human, a recording, and a fax machine. Depending on the criteria set, the telephone system then either drops the call or delivers it to the original calling device. This sequence is illustrated in Figure 4-11.

Figure 4-11. Make Predictive Call service example

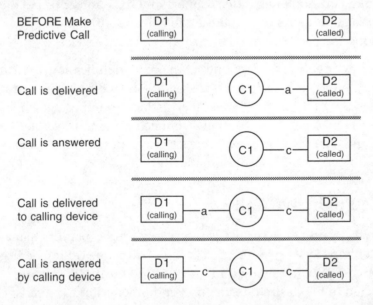

This service is even more powerful when the calling device is some type of routing device (such as an ACD, ACD group, or hunt group device). In the case of an ACD group, when the telephone system successfully places a call to the desired destination and then delivers the new call to the ACD group, it in turn redirects the call to the first available agent. This combination of the *make predictive call* service and an ACD group device allows a set of agents to maximize the time spent talking rather than waiting.

4.4.8. Last Number Dialed and Redial

The *get last number dialed* feature involves the logical device remembering the number of the called device for the last call it initiated. Combining this feature with the *make call* service is referred to as *redial*.

4.5. Call Associated Information

Whenever a new call is created, the telephone system begins associating various important pieces of control information with it. In the services we have explored so far, we have already seen that two key pieces of information associated with a new call are the called device and the calling device. Another key piece of information associated with external calls is the network interface device that is involved.

4.5.1. CallerID and Automatic Number Identification (ANI)

Aside from knowledge of a call's intended destination, the most important piece of information associated with a call is the identity of the calling device. This information is used both by the telephone network for billing purposes and by those being called.

Depending on the features of a particular telephone system and the features subscribed to from an applicable external telephone network, the feature that provides this information to the called device is referred to as either *callerID* or *automatic number identification (ANI)*.[4-9] (These specific features of the external telephone network, or *service offerings*, are described in Chapter 5.)

The called party in a call may use this information to determine how a call should be handled in an endless variety of money- and time-saving ways. Uses of this information include:

- Determining the name of the person calling by automatically looking up the number in a directory before answering;

- Deciding whether or not to accept or answer the call;

- Deciding what information should be automatically presented to the called party in order to best handle the call;

- Creating new database entries for tracking information provided by this caller;

- Logging information about the call (e.g., caller and length of call) for future reference or billing;

- Deciding if the call should be redirected and, if so, the destination; and

- Deciding what person is best able to answer the call.

Developing solutions that take advantage of callerID and ANI information is one of the most readily achievable ways to benefit from CTI technology.

Get and Set CallerID Status Services

The *Get CallerID Status* and *Set CallerID Status* services relate to a logical device element's ability to suppress the delivery of callerID information with calls that it places. If this feature is used to activate

---

4-9. ANI — ANI is slightly different from callerID in that it delivers the billing number of the calling device, that is, the calling number that is used for billing purposes. Typically the two numbers are the same. When they are not, it is usually because a shared network interface device is being used to place a call on the network but the call is being billed to a single main number. The actual device originating the call from the network's point of view is the network interface device, but in this case calls from any of the shared network interface devices are billed to a single main number; ANI will reflect the main number and thus will be different from (and more useful than) callerID.

CallerID blocking, then calling device information for calls made by the logical device element will be sent to the called device only in the case of toll-free calls and emergency calls (such as 1-800, 1-888, 1-900, and 911 in North America).

CallerID blocking may be *complete blocking* or *selective blocking*. With complete blocking, every new call made is blocked from sending callerID until the service is deactivated. With selective blocking, the callerID blocking only applies to the next call that is placed and must be reactivated for each subsequent call.

4.5.2. Dialed Number Identification Service (DNIS)

Depending on the features of a particular telephone system and the features subscribed to from an applicable external telephone network, *dialed number identification service (DNIS)* may be available. (DNIS service offerings are described in further detail in Chapter 5.)

With calls that are delivered to a particular destination device, dialed number identification service provides the actual number that was dialed by the calling device. This feature is used in scenarios where multiple numbers in one or more networks are all redirected such that calls to any of the numbers are routed to a single designated device.

One way of taking advantage of the DNIS feature is by centralizing all calls through a single routing device (typically an ACD, ACD group, or hunt group). In this way the utilization of people and telephony resources for handling calls can be maximized. The DNIS feature allows the telephone system to know the actual number dialed, even though all calls are being delivered to the same device. This information then can be used in much the same way that CallerID and ANI information is used. For example, you could subscribe to a set of numbers in the external network (such as 800-BUY-FOOD, 800-BUY-CARS, 800-BUY-HATS) and have each one directed to the same group of operators using your telephone system. They could then handle all of the calls to any of these numbers. Using DNIS information, the telephone system could:

- Direct the call to the best available operator for the appropriate product category;

- Inform the operator whether to say "Hello, Acme Grocery," "Hello, Cars-Are-Us," or "Hello, Mad Hatter's" when answering each call;

- Present the appropriate order form for the right sales activity; and

- Log the call activity in the appropriate database.

Developing solutions that take advantage of DNIS information is another excellent way to benefit from CTI technology.

4.5.3. Last Redirected Device

The *last redirected device* information associated with a call identifies the last device that rerouted this call. Like CallerID and DNIS information, this is very useful for interpreting why a particular call is being delivered to a particular device. In fact, the last redirected device actually is used, or interpreted, as a combination of called device and calling device. It is a device to which the call was previously directed, and it is the device that is responsible for redirecting the call to its current destination. Typical uses of this information include:

- Answering a call differently, depending on whether or not it was already answered by an operator or assistant.

- Identifying which of several forwarded calls came from a particular forwarder, independent of what number was originally dialed. For example, an assistant answering calls on behalf of two different managers could identify which manager forwarded the call.

- Allowing a voice mail system to play the message for the correct person when their call is redirected to voice mail because it was not answered.

- Determining whether or not to answer a call based on whether it has been screened by an appropriate person.

Support for the last redirected device information is another essential feature in any system.

4.5.4. Account and Authorization Codes

One feature associated with many systems is support for *account codes* and/or *authorization codes*. These are codes that are associated with the call in some way when it is created; these codes (in addition to information about the calling device) may be used for billing the call to the correct person, project, or organization. These pieces of control information also may be used for determining whether or not a call is external, what network interface device should be used, and which other call-related features might be applied.

4.5.5. Correlator Data

Correlator data is an arbitrary block of use-specific information that can be attached to a call. Correlator data is attached to calls in order to allow the many different devices that may interact with a call as it is routed through a telephone system to have access to some piece of common information, so that the caller need not provide it again and again.

Typical examples of the use of correlator data include:

- A customer's credit card number is attached to a call so that, as her call is transferred between different agents responsible for selling different products, she does not need to provide this information again and again.

- A man calls a local store, asking if a particular product is in stock. After describing the desired product to a clerk, the product's part number is determined and attached to the call. Unfortunately the store doesn't have it on hand, but the clerk transfers[4-10] the call to another store that is likely to have the product in stock. The clerk

193

at the second store does not need to repeat the interrogation process to determine the product number because it is attached to the call.

- A woman calls a travel agent about some vacation plans. The receptionist answers the call and discovers that the woman does not yet have an account. He then creates a new database entry for the woman and gathers all the pertinent information from her. The woman's newly created customer number is attached to the call as correlator data. All the travel planners in the office are busy with other callers, so the receptionist puts the call on hold[4-10] for the next available agent. The next person available retrieves[4-10] the call and, thanks to the correlator data attached, can immediately pull up her customer record. In fact, when it is discovered that the woman's travel plans involve South America, the travel planner transfers the call to the agency's expert on the subject and once again the woman's information travels with her call.

Correlator data allows information to travel with a telephone call. While it is useful by itself (as information displayed on a telephone display), it is very powerful, if not essential, when used in a CTI solution.

4.5.6. User Data

User data are blocks of information that are broadcast to all the devices in a call at any point in the life of a call. Its association with a call may or may not be in conjunction with a call control service. In any case, however, it is distinct from correlator data because it is a "one-shot" mechanism and the data is not persistent.

4-10. Transfer, Hold, and Retrieve switching services — The *transfer, hold,* and *retrieve* switching services is described fully later in this chapter. The *transfer* switching service involves moving a call to a different device. *Hold* and *retrieve* involve suspending and restoring the media stream associated with a particular connection.

Typical examples of user data usage include:

- Delivering VersitCard electronic business card information to other participants in a call;

- Providing vendor, customer, or account number information;

- Authenticating callers through password or cryptographic exchange; and

- Exchanging an encryption key to be used in a subsequent data transfer over the Internet.

Like correlator data, user data can be a very powerful feature in the implementation of a CTI solution.

4.6. External Incoming Calls

Support for *external incoming* calls is a feature that allows calls from an external network to be connected to devices inside a telephone system, using network interface devices. Network interface devices are the proxies within the telephone system that represent the calling device's participation in a call. This is shown in Figure 4-12. In this illustration, the "?" symbol indicates that because the connection between the calling device (D1) and the network interface device (D2) is entirely outside of the telephone system, the external network may or may not provide information about its state. External incoming calls are different from external outgoing calls in that the calling device is outside the telephone system for the incoming case, and is inside the system for the outgoing case.

Figure 4-12. External Incoming calls

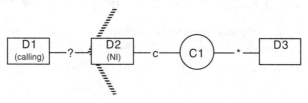

Once a network interface device has been associated with an external calling device, the new incoming call must be presented to a device inside the telephone system. The destination for the call is chosen in one of three basic ways, depending on the specified called device and the subscribed external network features:

- Fixed network interface device association
    All incoming calls on a specific network interface are connected to a predetermined device.

- Selectable network interface device association
    Each incoming call on a network interface indicates its own desired destination device.

- Attendant
    Incoming calls are presented to an attendant of some sort that provides assistance in connecting to the desired destination.

4.6.1. Fixed Network Interface Device Association

A telephone system that supports *fixed network interface device association* allows specific network interface devices to be configured so that, when a new external call is connected to them, they originate a call to a single specific device inside the telephone system. This feature is sometimes referred to as *DIL* or *direct-in-line*.

An example of fixed network interface association is shown in Figure 4-13. In this case, D1 places a call to the telephone system in question by dialing the telephone number in the external network (555-1234, for example) associated with network interface D2. In turn, D2 has a fixed association with device D3, so it delivers the call from D1 by originating a call to D3 inside the telephone system.

The relationship between the network interface device and its associated device also can be based on predetermined rules. For example, the association could be fixed with one device during the day and with a different device at night.

Figure 4-13. Fixed network interface device association

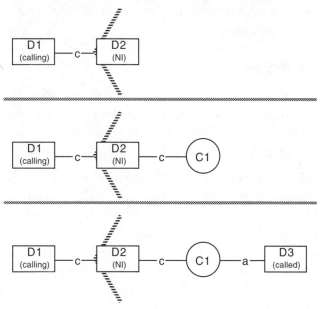

In any case, the device accepting the calls could be an individual station device, or it could be a routing device such as a hunt group, ACD group, or ACD that will, in turn, deliver a call to its destination.

4.6.2. Selectable Device Association

A telephone system that supports *selectable device association* allows each new external call to select its own destination within the telephone system by communicating this information when it connects to the network interface device.

There are three common forms of this functionality:

- Sub-addressing
- Direct Inward Dialing (DID)
- Direct Inward System Access (DISA)

In each case, information flows from the external network to the telephone system after a call has been presented to the network interface device involved.

Sub-addressing

Sub-addressing is a feature of certain telephone networks[4-11] that allows a sub-address to be dialed along with the telephone number of the network interface device. The sub-address effectively provides a "hint" as to which device in the telephone system is preferred.

The sub-address may be handled in different ways by the receiving telephone system, but the typical implementation is a simple variation on the fixed network interface device association. In this case, the network interface device that accepts the call has a fixed association with a bridged logical device element. The new call triggers all the bridged appearances to begin alerting; audible ringing is suppressed, however, for all physical devices other than the one indicated by the sub-address. Each device also can use the sub-address information to determine if it wants to answer or not. This approach is illustrated in Figure 4-14. Here L3 is the bridged logical device element, and P4 and P5 are the associated physical devices. If the sub-address dialed corresponds to P5, only it will ring.

Figure 4-14. Typical sub-addressing implementation

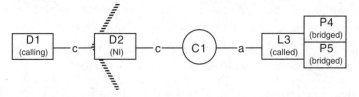

DID

Direct inward dialing (DID) is a very popular feature that involves assigning telephone numbers in the external network directly to selected devices inside the telephone system. These devices are then referred to as *DID extensions.*

When an external caller dials the telephone number corresponding to one of the DID extensions, the external network delivers the call to one of a set of "root" DID network interface devices. The external network then dials the last two, three, or four digits of the DID extension so that the network interface device can determine the appropriate destination device.

The sequence for DID may be implemented as shown for DIL in Figure 4-13 or as shown for DISA in Figure 4-15. In both cases, D1 places a call to the telephone system in question by dialing the telephone number in the external network for the particular DID extension (555-4220, for example). The network connects the call to an appropriate network interface device (555-4000 for purposes of this example) and then sends the digits "220". The network interface, D2 in this example, originates a call to D3 (which is identified as "N40220" inside the telephone system and has been assigned the number 555-4220 externally).

DISA

Direct inward system access (or *DISA*) is similar to DID but, rather than having the network specify the desired device, the caller dials the desired number directly. In fact, DISA can be much more than merely the ability to dial an extension. As the name implies, it represents complete access to the telephone system. Once a remote device connects to a DISA network interface device, the remote extension is treated as if it were a basic station device directly attached to the telephone system. Any commands that such a device can send to the telephone system can be issued by the external device, subject to class-of-service restrictions.

An example of a DISA sequence is shown in Figure 4-15. In this example, D1 places a call to the telephone system in question by dialing the telephone number ("555-4444", for example) in the external network for the special DISA network interface device. This DISA device, D2 in this example, then initiates a new call in the system. D1 will then hear dial tone just as if it were a station device that had just gone off-hook. In this example, D1 then dials the appropriate number for D3 ("40220", for example) and just as if it were a locally dialed call, the new call is originated and delivered to device D3.

Figure 4-15. Direct inward system access (DISA)

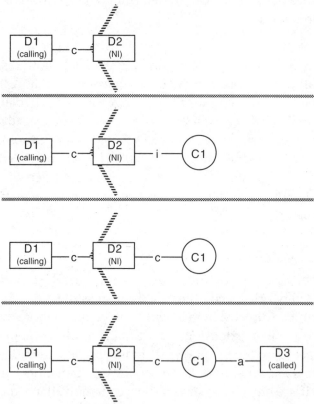

DISA potentially provides external callers with access to every feature and capability of your telephone system. As a result, the DISA feature is a serious source of toll fraud problems. (See the sidebar "Toll Fraud.")

Toll Fraud

Toll fraud refers to the misuse, or theft, of telephone services.

Telephone services are either billed implicitly, based on the calling device's network address, or explicitly, based on billing information (such as credit card number) provided during a call.

Toll fraud can therefore take place in two basic ways:

- Theft of billing codes
- Unauthorized access to facilities for placing long-distance calls

Billing codes (like calling card numbers) are primarily stolen by so-called "shoulder-surfers" who look over your shoulder as you punch in your billing information. They are also stolen by hackers breaking into phone company computer records, and by employees who steal and sell this information.

The principal way to steal access to a telephone system is through the DISA feature. Unless restrictions are placed on the features to which a DISA user has access, someone can call a DISA number and then, with full access to all of a telephone system's functionality, can place long-distance calls that will be billed to the owner of the telephone system. This means that if you are implementing support for DISA in your telephone system, you should make sure that you appropriately restrict the class of service that applies to it. Another way to secure DISA is to use callerID to restrict use to only specifically identified external devices. Many companies are switching to auto-attendants as an alternative to DISA for more control and security.

The latest form of toll fraud involves thieves who monitor cellular telephone traffic and intercept the code that uniquely identifies your cellular phone. They then program their cellular phone to be a clone of yours, and they can bill their calls to you because their phone has become almost indistinguishable from yours.

4.6.3. Attendant

The traditional approach for routing telephone calls between one system and another has involved using a human being who, with access to appropriate controls, can route calls manually. These people are traditionally called *telephone operators* or just *operators*. The term *attendant* refers to a designated operator (or one of a group of operators) for a private telephone system.

The attendant case is basically the same as the fixed network device association case except that the caller cannot directly dial the desired device. If the attendant feature applies to a particular network interface device (the default case for most systems), all calls arriving at that network interface are delivered to the designated attendant device or attendant group device.

The use of an attendant to complete delivery of a call is shown in Figure 4-16.

Attendant Consoles

The station device designated as the attendant for handling incoming calls (and other special attendant-related features) is referred to as the *attendant console*. (They used to be called *switchboards* when attendants were called operators.) The rate of call delivery to an attendant console typically is quite high, so they generally are designed and configured as multi-appearance devices. (See Chapter 5, section 5.6.7 for one example of an attendant console.)

Auto-attendant

Auto-attendants are a popular alternative to the traditional human attendant. An auto-attendant is a device that automatically answers external incoming calls, plays an appropriate greeting, and requests that the caller choose from a specific list of options where the call is to be directed. Depending on the implementation, the auto-attendant may allow the caller to specify the actual extension desired, allow the caller to dial the desired extension using the person's name, or look up the desired person in an interactive directory.

Figure 4-16. Attendant operation

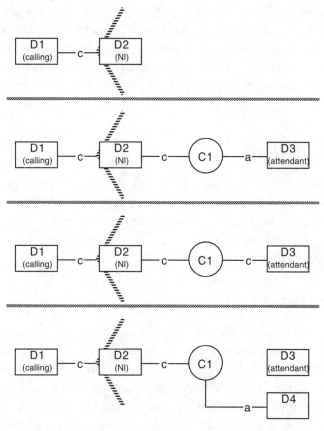

Auto-attendants provide a more secure and user-friendly "self-service" interface than DISA, so many people are switching from DISA to auto-attendants. Another benefit of auto-attendants is that they can be programmed to automatically handle fax and modem calls correctly, so that dedicated network interface devices are not required.

For most small telephone system owners, who generally do not have human attendants due to the obvious expense, auto-attendants are an extremely cost-effective alternative. It might even be desirable for a telephone system with only a single network interface device to have an automated attendant to screen calls.

Building auto-attendants is one of the fastest growing areas of CTI solution development. This is another area where the power and benefits of CTI technology are evident. CTI technology allows for the customization and day-to-day optimization of a system's automated attendant. The addition of technology such as text-to-speech and speech recognition allow for very simple and compelling interactions with callers.

4.7. Call Routing

Routing refers to all the ways that a call progresses through a telephone system. Each step in a routing sequence involves presenting a call to a device and having it either answer the call or direct it elsewhere. As long as a call exists, every new destination it reaches becomes part of its routing history.

4.7.1. Do Not Disturb

A feature called *do not disturb* can prevent the routing process from even starting at a particular device. If the do not disturb feature is activated for a particular logical device element, then all calls made to the device that satisfy specified criteria are rejected before they are ever presented to the destination. The rejected call may either fail or be redirected elsewhere. Do not disturb is illustrated in Figure 4-17. In this example D2 has activated the do not disturb feature; when D1 places a call to D2, the connection D2C1 immediately transitions to the *fail* state.

Figure 4-17. Do Not Disturb

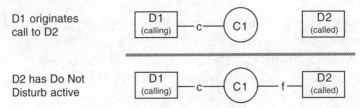

The do not disturb feature is used frequently in conjunction with the call forwarding feature (described later in this chapter), which allows the call to be redirected instead of transitioning to the *fail* state. This feature is very popular in hotel solutions, where guests do not want to be awakened by the telephone.

The do not disturb feature is managed with the *get do not disturb* and *set do not disturb* services. The feature can be activated to apply to all calls, or just those satisfying certain criteria relative to the origin of the call.

Internal / External

One option of the do not disturb feature is the ability to specify that the feature is to apply only to calls originating inside or outside the telephone system, rather than to all calls.

Selective Blocking

Selective blocking is a particular option of the do not disturb feature that allows one or more specific devices to be blacklisted. Calls made from these devices, and only these devices, will be rejected or redirected before they are presented to the destination.

Selective blocking is an example of a feature that is made largely obsolete by CTI. With CTI technology, call blocking can be implemented on your own computer system by setting it up to drop calls from blacklisted callers immediately. Unlike the telephony service, however a CTI-based implementation would have virtually no limit to the number of blocked callers and could be endlessly customized. For example, it could log attempts by blocked callers before rejecting their calls, it could block different callers at different times, it could play a prerecorded message to blocked callers before dropping their calls, etc.

4.7.2. Alerting

When call processing attempts to connect a device to a call, the connection between the call and the new device is in the *alerting* state. There are three modes of the *alerting* state through which a connection may transition during the process of attempting to establish a connection. A telephone system implementation may support one or more modes for a given device. The modes of *alerting*, in the order through which they transition, are the following:

1. *Entering distribution* mode

2. *Offered* mode

3. *Ringing* mode

The caller hears ringback as long as the call is in the *ringing* mode of the *alerting* state. Whether or not the caller hears ringback in the other modes depends upon the telephone system and/or the external network.

The first mode of the *alerting* state, *entering distribution*, refers to a state in which the call is being associated with a device with the specific intent that it be routed elsewhere. This is the mode normally used when a call is presented to a device for routing purposes. The use of the offered and ringing modes are described later in this chapter.

4.7.3. Queuing

Another connection state often observed in routing sequences is the *queued* state. This state reflects the fact that call progress has been suspended but not stalled. Typically connections are transitioned to the *queued* state while waiting for some resource to become available or some routing decision to be made.

Most implementations do have timers associated with *queued* calls to ensure that a call is not *queued* indefinitely; these timers are much longer than those for states such as *alerting*, however.

4.7.4. ACD Features

In many telephone systems, the first device that an incoming call is presented to is an ACD device. For example, auto-attendants that are built into the telephone system are ACDs.

An ACD device has built-in rules for distributing calls and may make decisions based on any of the call control information discussed earlier, as well as the time of day, the last device the call was redirected from, and information that it captures directly from the caller (through DTMF digit detection, for example).

ACDs that have *visible ACD-related devices* use independent devices (with separate connections to the call). A sequence using ACDs with this model is shown in Figure 4-18. In this example, the ACD device D2 is programmed to queue its connection if it can't immediately find an appropriate destination. It then enlists the help of a second ACD device, D4, which is also queued to the call. A media access device, D3, is connected to the call to play music to the caller while waiting for an available destination. When an appropriate destination (D5) is found, the "helper" devices are dropped from the call and it is diverted to D5.

With ACDs that have *non-visible ACD-related devices*, on the other hand, all of the media resources (and other resources) that they rely upon are internal to the ACD device itself. Using ACDs with this model, the same ACD activity appears as the sequence shown in Figure 4-19.

4.7.5. ACD Group and Hunt Group Features and Services

ACD groups and hunt groups behave in a fashion identical to that described for ACDs. The only difference is that these devices can only distribute calls to a finite set of devices. In the case of hunt groups, the group of devices is fixed and the only criterion for presentation of new calls is availability. In the case of ACD groups, the group consists of logged-on agents and the status of an agent is the basis for presenting a call.

Figure 4-18. Visible ACD-related devices model

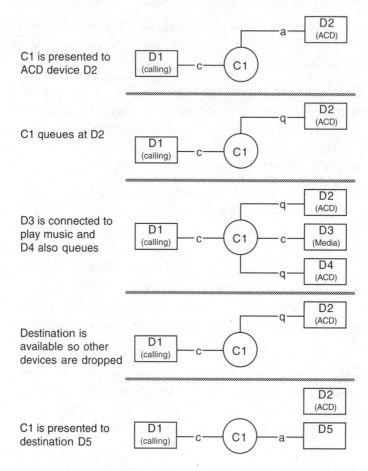

Get and Set Agent Status

The *set agent status* service is used by a device to log on to a particular ACD group, log off again, or indicate some other agent status during the cycle of handling calls presented by the ACD group. The ACD group will present a new call to a device only if its corresponding agent status is *agent ready*. If there are no agents with this status, the ACD group will queue the call. The *get agent status* service can be used by a device to observe the status of the agent association.

Figure 4-19. Non-visible ACD model

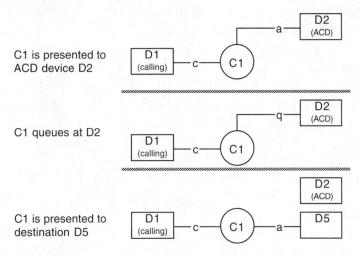

C1 is presented to
ACD device D2

C1 queues at D2

C1 is presented to
destination D5

4.7.6. Parking and Picking

Another service that involves queuing calls as part of the routing
process is the *park call* service. This service allows a call to be *parked*, or
queued, at a particular device. The *park-to* device may be a station
device or a park device from which the call can later be *picked* using the
directed pickup call service. The *park call* service is shown in Figure 4-20.

Figure 4-20. Park Call service

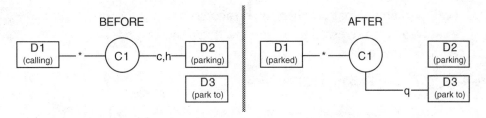

BEFORE   AFTER

Park to a Station

The most common use of parking involves an attendant queuing a call
on behalf of someone. Figure 4-21 shows an example that might be
found in a hotel. The hotel's attendant, D2, answers a call from

someone, D1, who wants the occupant of room 3002. The device in that room, D3, is busy; the caller indicates that she wants to wait, however, so the attendant parks the call.

Figure 4-21. Park Call to a station

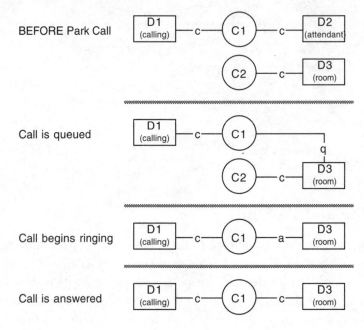

Park to a Park Device

A call parked at a park device can be accessed using the *directed pickup call* service (also referred to by some as the *unpark* or *retrieve park* service in this context). It is shown in Figure 4-22.

Figure 4-22. Directed Pickup Call service

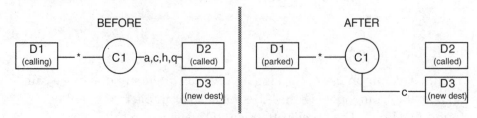

Figure 4-23 shows an example that might be found in an environment where people are not in fixed locations and do not have assigned telephones. In this example, the caller D1 asks for Fred Jones, who is roving somewhere in a large warehouse. The attendant, D2, parks the call to a special park device, D3. The attendant then makes an announcement on the public address system for Fred Jones to pick up the call on D3. Fred finds the nearest station, D4, and uses the directed pickup feature to connect to the call.

Figure 4-23. Parking and picking a call

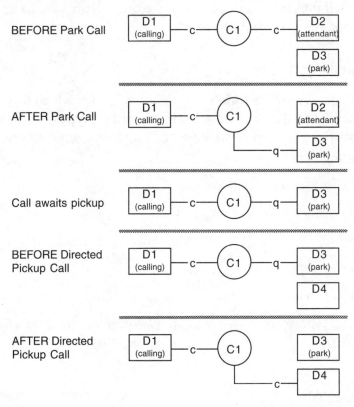

211

4.8. Forwarding and Coverage

Forwarding is a very popular telephony feature. It allows rules to be established for the automatic redirection of calls associated with specific logical device elements.

The forwarding feature is managed using the *set forwarding* and *get forwarding* services. Activating a particular forwarding rule involves specifying that the combination of a particular forwarding rule and a particular corresponding *forward-to* destination device should be activated. If the rule is then satisfied, the call is redirected to the corresponding forward destination.

4.8.1. Forwarding Types

Forwarding types are the different cases, or rules, for which forwarding can be activated. There are four basic forwarding types and three different *origination types* that allow for a total of twelve different forwarding types.

The basic forwarding types are:

- Immediate

- Busy

- No Answer

- Do Not Disturb

Origination type refers to where a call was originated. It may be one of the following:

- Internal

- External

- All

Immediate Forwarding

If *immediate forwarding* is active for a particular logical device element, then calls to that device (given the appropriate origination type) are immediately redirected to the corresponding forward-to destination. This is illustrated in Figure 4-24.

Figure 4-24. Immediate forwarding

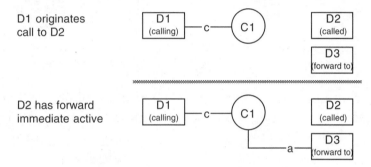

Immediate forwarding typically is activated when users will be away from their phones for a period of time and would like to have calls redirected, either to a temporary location (such as someone else's office or a cellular phone) or to someone who will be covering for them.

Busy Forwarding

If *busy forwarding* is active for a particular logical device element, then calls intended for that device while it is busy and therefore incapable of accepting calls (given the appropriate origination type) are forwarded to the corresponding forward-to destination. *Busy* forwarding is illustrated in Figure 4-25.

This type of forwarding is occasionally also referred to as *roll-over*. A call is said to "roll over to coverage" when this rule is triggered. The term *coverage* refers to the device responsible for covering for the device being called in the event that a call is not answered.

Figure 4-25. Busy forwarding

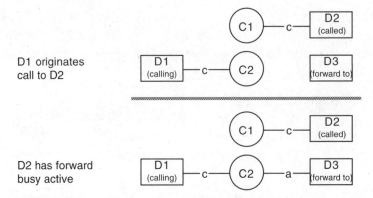

D1 originates
call to D2

D2 has forward
busy active

Busy forwarding is typically activated all the time for people that have voice mail, an answering service, or a full-time assistant that can handle calls that overflow.

No Answer Forwarding

In the case of *no answer forwarding*, the rule includes the number of rings to wait before giving up. If a call is presented (in the *ringing* mode of the *alerting* state) to a logical device with this type of forwarding active, it remains associated with the device for the number of rings specified. If it has not been answered, it is redirected to the forward-to device specified. This feature is illustrated in Figure 4-26.

No answer forwarding, like *busy* forwarding, is typically activated at all times for those with voice mail or an answering service.

Do Not Disturb Forwarding

The trigger for *do not disturb forwarding* is that the logical device element in question has activated the do not disturb feature and a call is about to be rejected as a result. In this case, the call is redirected to the specified forward-to device rather than being transitioned to the *fail* state. Contrast this behavior, shown in Figure 4-27, with the basic do not disturb behavior in Figure 4-17.

Figure 4-26. No answer forwarding

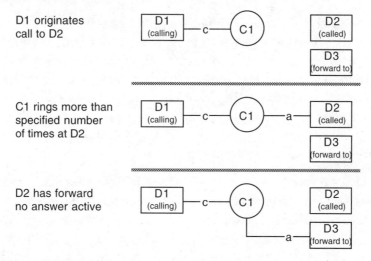

D1 originates
call to D2

C1 rings more than
specified number
of times at D2

D2 has forward
no answer active

Figure 4-27. Do not disturb forwarding

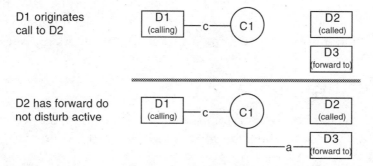

D1 originates
call to D2

D2 has forward do
not disturb active

Do not disturb forwarding never triggers unless the do not disturb feature is active, so the latter acts as a switch to turn the actual forwarding rules on and off. As a result, *do not disturb* forwarding typically is activated at all times for those with voice mail or an answering service. When users wish not to be disturbed, they activate the do not disturb feature, which in turn activates forwarding. This saves having to activate the forwarding each time.

4.8.2. System Default and User Specified Forwarding

There are often a number of default settings associated with implementations of the forwarding feature. A telephone system may have a set of *system default* forwarding rules for each logical device element in the system. This is often the case because the settings for *busy*, *no answer*, and *do not disturb* are typically all the same and all typically set to forward to a voice mail system if one exists. To eliminate the need to set all of these rules individually for each device, they are usually set up as system defaults.

Each device then can be managed individually to override these defaults, or turn them off altogether, on an as-needed basis with *user specified* forwarding. User specified forwarding also has associated defaults. If the *set forwarding* service is used, and a forward-to device is specified but a forwarding type is not, the telephone system will assume that its default forwarding type (normally *immediate–all*) should be used. If the *set forwarding* service is used and a forwarding type is specified without a forward-to device, the default forward destination (normally a voice mail system or the attendant) is used.

4.9. Offering

Once a call has been routed to its intended destination, the second mode of the *alerting* state, the *offered* mode, may come into play.

With the *offering* feature, the telephone system presents a new call in offered mode to a device in order to give the device the opportunity to accept, reject, or deflect the call before it transitions to the *ringing* mode. This gives the device the opportunity to screen calls before a human being is given the opportunity to interact with them.

If no action is taken, the *offered* mode times out and transitions automatically to the *ringing* mode. Depending on the implementation of the offering feature in a particular telephone system, the device may be able to answer the call directly while it is still in the *offered* mode.

4.9.1. Accepting

The *accept call* service is used to accept calls in the *offered* mode of the *alerting* state. When the service completes, the specified connection has transitioned from the *offered* mode to the *ringing* mode. *Accept call* is shown graphically in Figure 4-28.

Figure 4-28. Accept Call service

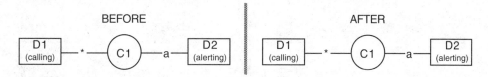

4.9.2. Deflecting

Another option for dealing with a call that is presented in the *offered* mode is to *deflect* it using the *deflect call* service.

The *deflect call* service is not limited to the *offering* case. It can operate on any call that is *alerting* or *queuing* at a device and divert it to a new destination device. If the service completes successfully, there is no connection at the deflecting device and the connection at the new destination device is one of the following:

- *Alerting* – the call has already reached the new destination device and the telephone system is attempting to connect the call.

- *Connected* – if the new destination is on an external network, this indicates that the call is connected to a network interface device; otherwise, the call not only reached the called device but has already been answered.

- *Fail* – call processing stalled while trying to connect to the new destination device, most likely because it was busy.

- *Queued* – the call is being queued at the called device, probably because it is some type of distribution device.

- *Null* – the call was redirected once again, so there is no connection to the new call at the destination device; there probably is a connection to the new call at some other device, however.

This service is described graphically in Figure 4-29.

Figure 4-29. Deflect Call service

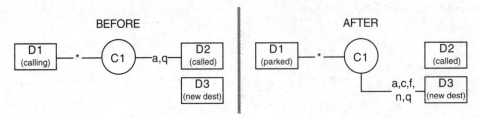

An example of an offering sequence in which the *deflect call* service is used appears in Figure 4-30. In this example, D1 places call C1 to device D2. The call is presented in the *offered* mode and D2 decides that it would be better handled by D3, so the *deflect call* service is used.

Figure 4-30. Deflect Call service in offered scenario

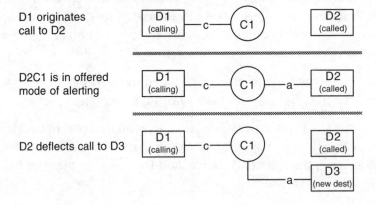

4.9.3. Rejecting

A device may reject an offered call. This involves simply dropping the call using the *clear connection* service.

Rejecting a call in the *offered* mode allows a device to implement selective blocking locally. Rather than rely on the telephone system or the telephone network to provide selective blocking, the device itself checks the information associated with a particular call against its list and drops the call before it transitions to the ringing state.

4.10. Answering

In section 4.3.2 we looked at the basic implementation of the *answer call* service. Answering a call involves making the transition from the *alerting* (or *queued*) state to the *connected* state. In this section we'll look at other ways that calls can be answered.

4.10.1. Auto Answer

The *set auto answer* service instructs a device to answer calls automatically after a certain number of rings. When the a connection to the device has been in the *ringing* mode of the *alerting* state for the appropriate number of rings, the system automatically answers it as if an *answer call* service had been used. If the number of rings is zero, the call is auto answered the instant it is presented to the device.

This service typically is used by voice mail systems, auto-attendants, and fax machines that operate in an autonomous fashion.

4.10.2. Pickup

Another way that calls may be answered is through the *pickup* services. These services allow a device to answer a call that is at a different device.

Directed Pickup Call

In section 4.7.6 we saw how the *directed pickup call* service is used in conjunction with the *park call* service to park and pick calls. This service also can be used to answer a call that is *alerting* at a different device. This use of the *directed pickup call* service is also referred to as *dial pickup* and *reverse transfer*.

Directed pickup call typically is used when a person hears someone else's phone ringing and wants to answer it. The most common situation where this occurs is in an office building after hours. There is no attendant to answer incoming calls, so people working late can use *directed pickup call* to answer the calls ringing at the attendant console.[4-12]

An example of the use of *directed pickup call* is shown in Figure 4-31. Note that this sequence is very similar to the sequence for *deflect call* in Figure 4-30. The key difference is that the device picking the call is actually answering it, so when the service completes, D3 is in the *connected* state.

Figure 4-31. Directed Pickup Call for an alerting device

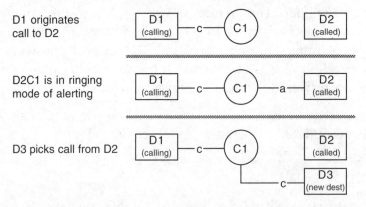

| D1 originates call to D2 | D2C1 is in ringing mode of alerting | D3 picks call from D2 |

4-12. Night bell — Some installations have an especially loud ringer, referred to as the *night bell*, that is mounted in a central place in the office. The night bell can be activated in the evening so that the presence of a pickable call is heard throughout the office.

Group Pickup Call

The *group pickup call* service is very similar to the *directed pickup call* service. Rather than specifying a particular device from which to pick the call to answer, however, the call is picked from another device by specifying the pick group device with which it is associated. By default, the pick group device used is the one with which the new destination device is associated. The *group pickup call* service finds an appropriate call associated with one of the devices in the group and redirects it to the new destination device. This is illustrated in Figure 4-32.

Figure 4-32. Group Pickup Call service

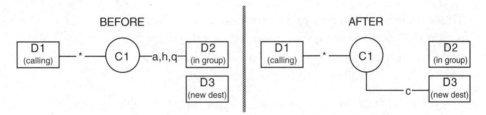

4.11. Suspending Calls

Once a connection has been answered and is in the *connected* state, the media stream(s) associated with the call flow through the established media stream channels. This continues until the call is placed into the *hold* state or is cleared.

When a connection is in the *hold* state, the media stream(s) associated with it are suspended and the media stream channels for it usually are deallocated. Generally there are six reasons to place a connection in the *hold* state.

1. As an alternative to muting the speaker and microphone.

2. To allow the call to be picked up somewhere else (using pickup services or by taking advantage of a bridged device configuration).

3. To reallocate the media stream channels being used in order to issue a command to the telephone system (if applicable for a given system).

4. To reallocate the media stream channels being used in order to participate in an additional call without dropping the first.

5. To reallocate the media stream channels being used in order to contact a second device to which the first call is to be redirected (*transferred*).

6. To reallocate the media stream channels being used in order to create a second call with which the first is to be joined (*conferenced*).

The term *hard hold* refers to an implementation of the *hold* state in which the media stream is suspended but the corresponding media stream channels cannot be reallocated for some other purpose. A hard hold supports the first two of the uses above.

The term *soft hold* refers to an implementation of the *hold* state in which the corresponding media stream channels can be reallocated for some other purpose. A soft hold is able to support all of the uses above, but some telephone systems require that the purpose for the *hold* state be specified in advance and then used only for the specified purpose.

Some telephone systems support optional *channel reservation*. If this feature is invoked when a connection is placed in the *hold* state, the associated media stream channels are not deallocated. If additional media stream channels are available, they can be used to allow for a soft hold; otherwise the result is a hard hold.

4.11.1. Hold

The *hold call* service places an active (e.g., *connected*) connection into the *hold* state. This single transition is the only stipulation of the completion criteria for this service. The *hold call* service is shown graphically in Figure 4-33.

Figure 4-33. Hold Call service

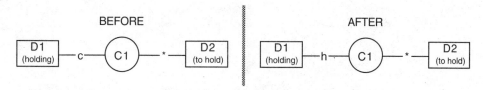

By default, the *hold call* service results in a hard hold, but many telephone system implementations attempt to provide a soft hold if possible. The sequence for a typical soft hold is illustrated in Figure 4-34. When call C1 is placed on hold by D1, the telephone system automatically creates a new call, C2, and reuses the media stream channels previously used by connection D1C1 for connection D1C2.

Figure 4-34. Soft hold implementation

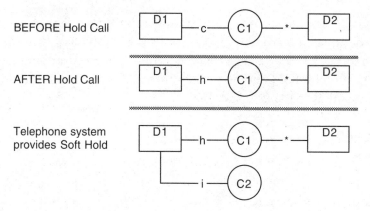

4.11.2. Consult

The *consultation call* service places an active (e.g., *connected*) connection into the *hold* state and originates a new call to specified destination. The name derives from the fact that this service typically is used to allow one person to leave a call and consult with someone else. This is shown graphically in Figure 4-35.

Figure 4-35. Consultation Call service

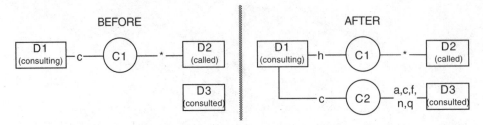

With respect to this service's ability to originate a new call, it operates in a fashion identical to the *make call* service. Therefore, in addition to placing the connection D1C1 into the *hold* state, the completion criteria are identical to those for *make call*. Multi-stage dialing and external outgoing calls are supported in the same way. Invoking the *consultation call* service without providing any digits to dial (thus initiating multi-stage dialing) is effectively a soft hold.

The *consultation call* service sometimes is referred to as a *compound service* because in many implementations it is equivalent to a a *hold call* service followed by a *make call* service. This is true for implementations that support soft hold and do not require any indication of the purpose for a transition to the *hold* state.

Consult Purpose

Consultation call is distinct from a *hold call / make call* combination in implementations where the *hold call* service only applies to the first two or three uses of hold listed earlier. With these telephone systems, *consultation call* is intended specifically for reallocating the media stream channels used for the first call in order to perform one of the last three uses listed. In addition, these implementations typically require that the *consult purpose* be specified. Consult purpose is one of:

- *consult only*;

- *transfer*;

- *conference*; or

- *conference and transfer*.

Each of these consult purposes can be thought of as a different variation of the service itself. Telephone systems of this variety typically have buttons on their telephone sets that correspond to these different variations of the *consultation call* service, and the individual services have names such as *setup transfer* and *setup conference*.

4.11.3. Retrieve

The *retrieve call* service, also referred to by some as the *unhold* service, transitions a connection from the *hold* state into the *connected* state. This single transition is the only stipulation of the completion criteria for this service. The *retrieve call* service is shown graphically in Figure 4-36.

Figure 4-36. Retrieve Call service

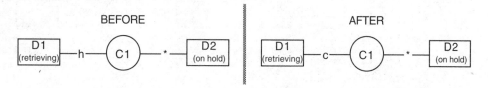

It is important to note that there are two common reasons for the *retrieve call* service to be unsuccessful:

1. An active connection for a voice call already is present at the physical device element represented by D1. Physical device elements can interact with only one active voice call at a time, so if an attempt is made to retrieve a call while another voice call is active, the service will not be successful.

2. No media stream channels are available to allocate to the connection. Even if there is no active voice connection tying up the physical device element concerned, the media stream channels needed to retrieve the previously held connection may have been used for digital data connections or for other physical devices in an independent–shared–bridged configuration. This situation can be avoided by invoking the channel reservation feature to ensure that the connection will be retrievable.

The *retrieve call* service is frequently used in bridged configurations where a call is placed on hold at one physical device element and is retrieved from a different one.

4.11.4. Alternate

The *alternate call* service operates on two connections at the same time. It takes one connection in the *connected* state and transitions it to the *hold* state. It transitions the other connection from the *hold* state to the *connected* state. The *alternate call* service is shown graphically in Figure 4-37.

Figure 4-37. Alternate Call service

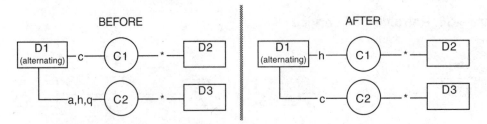

The *alternate call* service sometimes is referred to as a compound service because it is functionally equivalent to a *hold call* service followed by a *retrieve call* service.

The *alternate call* service typically is used in conjunction with the *consultation call* service. After consulting on the second call, *alternate call* can be used to toggle back and forth between the two calls.

4.11.5. Reconnect

The *reconnect call* service is similar to the *retrieve call* service except that it typically ensures that the necessary media stream channels will be available to retrieve a connection previously placed on *hold* by clearing another connection. The *reconnect call* service is shown graphically in Figure 4-38.

Figure 4-38. Reconnect Call service

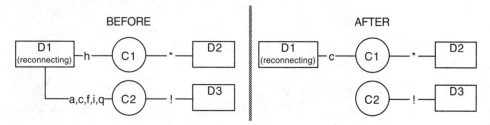

The *reconnect call* service typically is used in conjunction with the *consultation call* service. After consulting on the second call, *reconnect call* can be used to return to the first call in one step.

4.12. Transfer

Transferring (or *extending* as it is sometimes known) a call to a new destination is different from deflecting or forwarding a call because it involves a connection that has been answered. Deflecting and forwarding manipulate the routing of a call before any media stream channels have been allocated for the call at a given destination.

4.12.1. Transfer with Consult

A *two-step transfer* operation involves putting the call to be transferred on hold and consulting with the intended recipient of a call (typically using the *consultation call* service) and then executing the *transfer call* service. This service drops the transferring device from both calls and connects the other device(s) from the first call and the device(s) from the second call into a single new call. The *transfer call* service is shown graphically in Figure 4-39.

Figure 4-40 illustrates a typical sequence in which the *transfer call* service is used.

The *transfer call* service may be unsuccessful if the telephone system requires that the intent to transfer be indicated during the first step in the process and a consult purpose of *transfer* or *conference and transfer* was not provided. (Refer to consult purpose in section 4.11.2.)

Figure 4-39. Transfer Call service

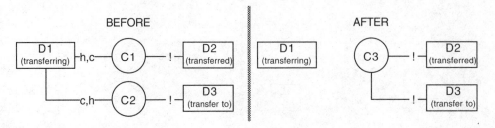

Figure 4-40. Two-step transfer call sequence

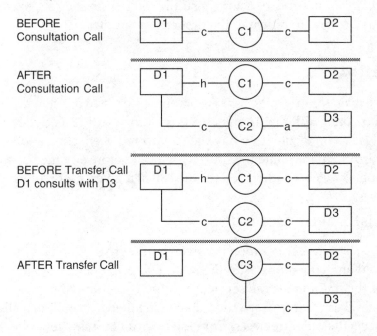

4.12.2. Single Step Transfer

Single step transfer call, also referred to as *blind transfer* by some, is similar to *transfer call* but it skips the consultation step. The *single step transfer call* service is shown graphically in Figure 4-41.

In one step this service drops D1 from the call it is transferring, C1, places a new call, C3, to the transferred-to device D3, and merges the remaining device(s) from call C1 into C3. At the completion of this

Figure 4-41. Single Step Transfer Call service

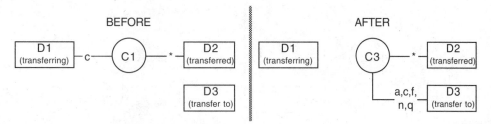

service, the result is very much as if D2 had used the *make call* service to call D3 directly. The state of connection D3C3 is the same as described for the called connection after successful completion of a *make call* service.

4.13. Multi-party Calls

All the services we have looked at so far have involved the creation and manipulation of two-device[4-13] calls. In this section we look at the services that allow the addition of multiple devices into calls to form multi-party calls.

4.13.1. Conference

The *conference call* service is very similar to the *transfer call* service in that it involves a two-step process that merges an original call with a second call to form a new call. The difference is that the conferencing device remains in the resulting call along with all of the devices involved in both the calls, so the result is a three-way (or more) call. The *conference call* service is shown graphically in Figure 4-42.

4-13. Two-device calls — Calls involving only two devices represent the majority of all calls, and they are frequently referred to by a number of different names. These include: *point-to-point calls*, *two-party calls*, *basic calls*, and *simple calls*.

Figure 4-42. Conference Call service

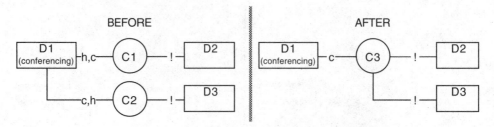

Typically the *consultation call* service is used to set up the second call; the *alternate call* service also may be used to toggle between the calls before actually conferencing them together. Figure 4-43 shows an example of a conference call sequence.

Figure 4-43. Two step conference call sequence

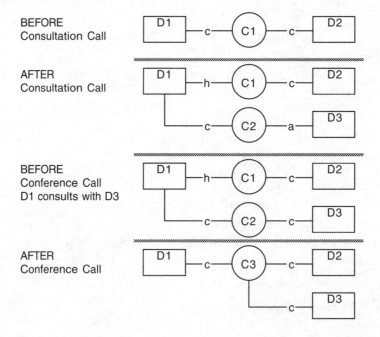

Like *transfer call*, the *conference call* service may be unsuccessful if the telephone system requires that the intent to conference be indicated during the first step in the process and a consult purpose for conferencing was not provided. (Refer to consult purpose in section 4.11.2.)

4.13.2. Single Step Conference

The *single step conference* collapses the two steps of the conference call process into one, as shown in Figure 4-44.

Figure 4-44. Single Step Conference Call service

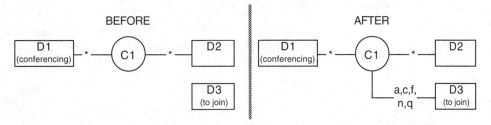

By specifying D3 as the destination for a single step conference involving call C1, the connection D3C1 is created in exactly the way that it would if any of the devices already in C1 had just placed a new call to D3 using the *make call* service. The difference is that all of the devices already in C1 remain in the call.

Traditionally single step conferencing has not been very popular because if the new connection to the call is not answered, all the participants in the call may hear endless ringback or busy tone. Unless the device to join is known to be expecting the call, or the telephone system provides support for the ability to drop an individual connection from a call (see *clear connection* in section 4.15.1), *single step conference call* is not a recommended service. On the other hand, if equipped with the ability to drop individual connections, *single step conference call* is a very powerful service because it allows a call to be established quickly with a group of selected devices.

4.13.3. Join

The *join call* service is very closely related to *single step conference call* but differs in one very important way. The *join call* service involves a device attempting to establish a new connection to an existing call, in contrast to the *single step conference call* service, which attempts to connect an existing call to a device. The *join call* service is shown graphically in Figure 4-45.

Figure 4-45. Join Call service

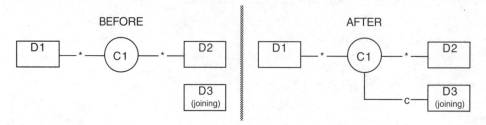

D3 uses the *join call* service to add itself to call C1. If the service is successful, the connection D3C1 is created in the *connected* state.

4.13.4. Silent Participation

Telephone systems may support the *silent participation* feature in conjunction with *single step conference call* and *join call*. Silent participation refers to creating the new connections as unidirectional, so that the added devices can listen into the call but cannot themselves be heard.

One situation where the silent participation feature is used with *single step conference call* is when one party in the call wants to record the conversation.

In another example, shown in Figure 4-46, D3 uses *join call* to add itself into call C1 with silent participation. D3 could be a supervisor who is quality-monitoring the calls being handled by a team of customer support agents. Using the *join call* service with silent participation allows the supervisor to listen discreetly into these calls in order to verify that customers are being treated in an appropriate fashion.

Figure 4-46. Join Call with silent participation

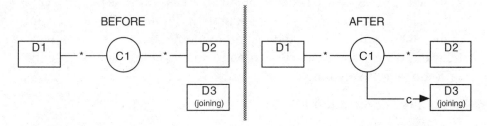

4.14. Call Failure

When call processing is unable to route a call and ultimately connect it to a destination for some reason, then call progress is said to be *stalled*. This section describes the ways that telephone systems handle calls in these situations, and the services that may be available when they occur.

4.14.1. Call Failure Handling

The *fail* state indicates that call progress was stalled for some reason and an attempt to associate a device and a call (or keep them associated) failed. The most common example of this state is attempting to connect to a device that is busy.

In most cases, all of the other active (e.g., *connected*) connections to the call will hear busy tone, or another appropriate failure tone, while a connection in the *fail* state is associated with the call.[4-14]

When a call ultimately stalls, the device for which it was intended might or might not (depending on the telephone system implementation) be aware of the connection attempt. Normally the connection is created and it transitions to the *fail* state. If the attempt to create the connection failed, however, there is no connection and the

---

4-14. Blocked — One case where the *fail* state is not associated with an audible tone is in the case where a bridged connection is blocked from a call. This is described in Chapter 3, section 3.9.3.

intended device will be unaware of the attempt. This is illustrated graphically in Figures 4-47 and 4-48. In both cases D1 has placed a call, C1, to device D2. In the first implementation (Figure 4-47), D2 is aware of the failed connection. In the second implementation (Figure 4-48), D2 is unaware of the connection attempt and its failure, in this case the call alone effectively failed.

Figure 4-47. Failed connection

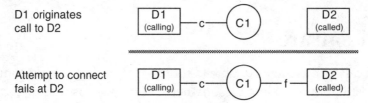

Figure 4-48. Failed connection attempt (failed call)

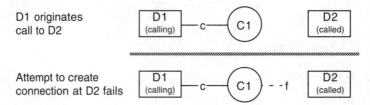

In both the scenarios illustrated, certain services still may be applied to the call (specifically to the connection D2C1 in the *fail* state in the first case, and to the *pseudo-failed* connection in the second case). One option is simply to drop the call or specific connections on the call (described in section 4.15). The other applicable services are called *call completion services* and are described in this section.

4.14.2. Camp On Call

One service that can be applied to a connection in the *fail* state is the *camp on call* service. This service queues a call that has failed for a given called device. The calling device remains connected to the call in the *connected* state and a connection in the *queued* state is established for the called device. Figure 4-49 shows the *camp on call* service in graphical form.

Figure 4-49. Camp On Call service

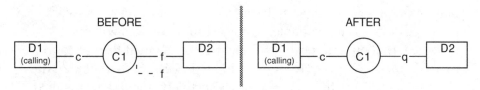

When the called device becomes available (and has dealt with any previously queued calls), the called device's connection transitions to the *alerting* state and it then may be answered by the called device.

An example showing how the *camp on call* feature works is illustrated in Figure 4-50. In this example, D1 places a call to D2 but the call fails because D2 is unavailable. D1 uses the *camp on call* service to queue the call at D2 (which results in the previously incomplete connection D2C1 being created in this case). When D2 becomes available, connection D2C1 transitions to the *alerting* state and D2 may answer it.

Figure 4-50. Camp On Call example

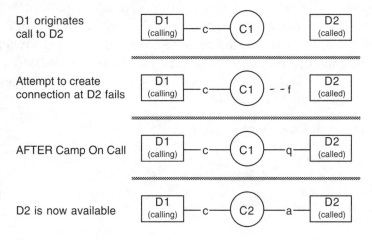

Typically this service, sometimes called *off-hook queuing*, is used when a call is placed and the call fails because the called device is busy. This feature is very useful not only for waiting on station device availability, however, but also for queuing to use network interface devices. When a large number of station devices share a small number of network interface devices, there will be times when no network interface devices are available. At these times, devices wishing to make external outgoing calls will see their calls fail at the appropriate network interface group (described in section 4.4.4). By invoking the *camp on call* service, these devices are effectively queued for the next available network interface device in the group. When a network interface device becomes available, the call proceeds.

4.14.3. Call Back

Another service that can be applied to a connection in the *fail* state is the *call back call-related* service, referred to as the *ring again* service by some. This service activates a feature that will cause the unavailable called device to be called again automatically when it becomes available.

The *call back call-related* service can be applied to the calling device's connection if the call itself failed, or if the connection to the called device is in the *fail*, *alerting*, *queued*, or *null* state. (It can be in the *null* state if the call was forwarded or deflected prior to failing.) When the service completes successfully, it clears the call and all its connections and leaves the call back feature set on the called device. This service is shown graphically in Figure 4-51.

Figure 4-51. Call Back Call-Related service

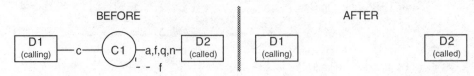

Figure 4-52 provides an example illustrating how the call back feature works. In this example, D1 places a call to D2 but the call fails because D2 is busy. D1 uses the *call back call-related* service, which clears the call and sets up the call back feature. When D2 becomes available, the call back feature is triggered and the telephone system attempts to place the appropriate call between D1 and D2. First it creates a new call, C2, with a corresponding connection to D1 and prompts D1 to go off-hook to accept it. (This is indicated by placing D1C2 in the *initiated* state.) When D1 goes off-hook, the call is originated to D2, so D1C2 transitions to *connected* and D2C2 transitions to *alerting*. (We know it will be *alerting* because it is available.)

Figure 4-52. Call Back example

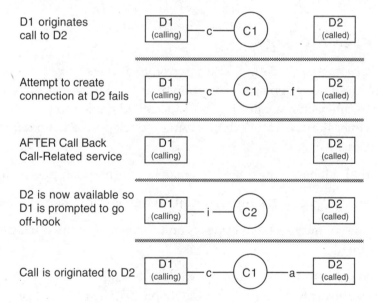

This feature is sometimes referred to as *on-hook queuing* because it allows a calling device to wait in queue for the called device to become available, without having to maintain a connection. In other words, it can wait for the call back while on-hook and not consume switching resources. Like *camp on call*, this service is typically used in conjunction with busy station devices, but it is also very good for queuing to use a network interface device. When a network interface device becomes available, the device is called back and the previously originated call can be attempted again. The trade-off between the two is the extra complexity of the call back feature versus the resource consumption of the camp on call feature.

Another variation of the call back feature is the *call back non-call-related* service. With the call-related version of the service, the device from which a call back is desired is indicated by the intended called device. If a particular device is known to be busy, however, the *call back non-call-related* service allows a call back to be registered against it without having to first place a call.

The *cancel call back* service allows one or all registered call backs to be cleared from the queue for a given device.

4.14.4. Call Back Message

The *call back message* feature, also referred to as the *call me message* feature on some systems, is a variation on the call back feature wherein a special call back indicator is set for the called device so that it can initiate the call back, rather than having the telephone system initiate the call back.

The actual call back "message" can take any form. Examples include:

- A special call back lamp that is lit on the physical element.

- A special message that appears on the display of the physical element.

- The name of the person who wants to be called (and the date and time they called) appears on the display of the physical element.

- A special tone is played just prior to dial tone when the handset goes off-hook.

In many implementations, a special "call back" button is provided on the physical device element, which triggers a *make call* service to place a call to the first device that registered an outstanding call back message.

The *call back message call-related* service allows the call back message feature to be set for the called device associated with a particular call. This service is otherwise identical to the *call back call-related* service (see Figure 4-51). The *call back message non-call-related* service allows the call back message feature to be set for any given device.

The *cancel call back message* service allows one or all registered call back messages to be cleared from the queue for a given device.

4.14.5. Intrude

Yet another way that failed calls may be handled is with the *intrude call* service. The *intrude call* service allows a device to "break into" another call in order to reach the called device if a call fails. There are two variations of the *intrude call* service, depending on the telephone system implementation. These are shown graphically in Figures 4-53 and 4-54. In both cases D1 placed a call to D2 that failed because D2 was busy in call C2 with D3.

Figure 4-53 shows the first case of the *intrude call* service, where the result is that D2's connection to its original call, D2C2, is placed on *hold* and connection D2C1 is transitioned directly into the *connected* state.

Figure 4-53. Intrude Call service (case 1)

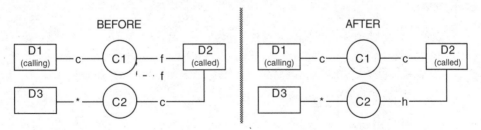

Figure 4-54 shows the second case of the *intrude call* service, where the result is that a new call C3 is created in which D1, D2, and D3 are all participants.

Figure 4-54. Intrude Call service (case 2)

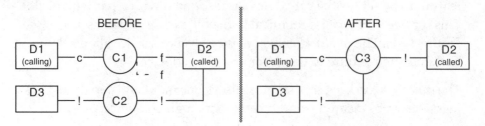

This second case also is referred to as *barge in* and the intruding device may have the option to join as a silent participant. This is illustrated in Figure 4-55. In this case, the intruding device can hear the conversation but cannot participate.

Figure 4-55. Intrude Call service (case 2) with silent participation

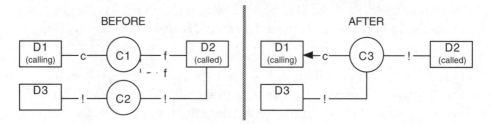

4.14.6. Recall

Telephone systems that support the *recall* feature have an alternative to failing calls in certain circumstances. It is a feature wherein a timer is associated with calls after certain services have been applied to them. The services that recall applies to include the following:

- Hold Call

- Transfer Call

- Single Step Transfer Call

- Deflect Call

- Park Call

If a device moves a call away, or places it on hold, using one of these services and that call fails to connect to a device before the associated timer expires, the recall feature is invoked and the call is returned to the device with a connection in the *alerting* state. Figure 4-56 illustrates an example of the recall feature in operation.

Figure 4-56. Recall feature

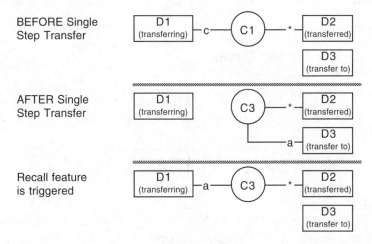

Recall is a very important feature because it ensures that callers are not "lost" or forgotten in the telephone system.

4.15. Dropping Calls and Participants

Now that we have looked at all of the ways to create, manipulate, and combine calls and their connections, we turn to how calls and connections are disposed of.

4.15.1. Clear Connection

The *clear connection* service either drops a specific connection from a call or inactivates a connection in a shared–bridged device configuration. The connection in question may be in any non-*null* state initially, and the state of all other connections in the call are not considered or directly affected by the *clear connection* service.

If the *clear connection* service completes successfully, the connection in question is transitioned to either the *null* state or the *queued* state. The *queued* state applies to a call at a device with shared–bridged appearances where at least one other appearance is participating in the same call. (This was illustrated in Figure 3-46 in Chapter 3, section 3.8.3.) In all other cases the connection will be completely disposed of with a transition to the *null* state.

The *clear connection* service is shown graphically in Figure 4-57.

Figure 4-57. Clear Connection service

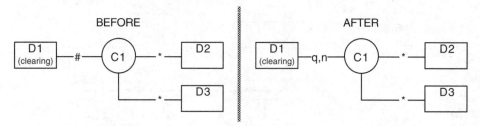

The connection being cleared may transition through the *fail* state on its way to *null* state under certain circumstances. Specifically, if a connection is cleared but the associated hookswitch on the physical element is off-hook, the connection will enter the *fail* state. It will remain in *fail* state until the physical element is set on-hook. When this occurs, the connection will transition to the *null* state and the service will have completed.

If there were only two devices with connections in the call prior to one of the connections being cleared, the call itself generally will be cleared. If the second device in the call was a network interface device, there may be multiple devices in the external network that remain in the call, even though the call no longer exists inside the telephone system and the network interface device is no longer being used.

The *clear connection* service is very useful in multi-party calls because it allows individual devices to be dropped from the call as needed. An example of an excellent CTI solution involves conferencing software that allows participants in a conference to see the names of all of the people involved in a conference call. The chairperson for the conference can use this software to add people to the conference (through the *single step conference call* service) and drop people (through the *clear connection* service) without affecting the others in the conference.

4.15.2. Clear Call

The *clear call* service tears down an entire call and all of its connections regardless of the state of any of its connections. The *clear call* service is shown graphically in Figure 4-58.

Figure 4-58. Clear Call service

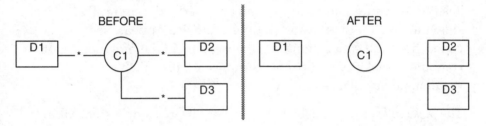

The *clear call* service is useful whenever it is desirable to shut down an entire multi-party call quickly and efficiently. Referring to the example of the conferencing software again, if the conference chairperson wanted to conclude the entire conference all at once, it would be quite painstaking if each individual participant had to be cleared one at a time using *clear connection*. The *clear call* service makes this a single step process.

4.16. Review

In this chapter we have seen the tremendous diversity of telephony functionality in the form of switching *services* and *features* that may be available in a telephone system.

This functionality ranges from the *basic* set of services of making, answering, and clearing calls (using the *make call*, *answer call*, and *clear connection* services), through to an extensive range of well-defined *supplementary* services. In addition, many telephone system vendors have implementations that include an *extended* set of functionality known as *vendor specific extensions*.

Telephony features and services can be described graphically by showing how the simple abstractions of device, appearance, call, connection, and connection state that we saw in the previous chapter combine to make even the most complicated functions easy to understand at a glance.

The *make call* service is used to originate calls in a single step. Multi-stage dialing is accomplished by combining it with the *dial digits* service. When a new call is created and its intended destination (the called device) has been identified, the call is connected to the calling device and is said to have been *originated*. If the calling device cannot go off-hook automatically, the telephone system may need to perform *prompting* to indicate that the device should be taken off-hook by a person. *Predictive dialing*, using the *predictive make call* service, allows the telephone system to place calls automatically and only uses a person's time if a call is successfully placed. A logical device that supports the *last number dialed* feature remembers the destination of the last originated call.

Calls may be *internal* to a telephone system, or they may be *external incoming* or *external outgoing*. External calls take advantage of network interface devices to connect to the telephone network outside a given telephone system. Special hunt group devices referred to as *network interface groups* (or *trunk groups*) route external outgoing calls to the next available network interface device. *Dial plan management* and *least cost routing* capabilities make telephone systems easier to use and cheaper to operate.

External incoming calls are routed to their intended destinations through one of three basic methods: fixed network interface device association (e.g., *DIL* service), selectable device association (e.g., *sub-addressing*, *DID* service, and the *DISA* feature), and through an attendant (e.g., a switchboard operator or an auto-attendant).

Associated with every call are critical pieces of information that can be put to good use by people and CTI systems. They include *CallerID* and *ANI*, *DNIS*, the *last redirected device*, *account* and *authorization codes*, *correlator data*, and *user data*. A range of services, such as *Get CallerID* and *Set CallerID* interact with the features that manage call information.

Calls may be delivered directly to a special routing device (such as an ACD, ACD group, or hunt group) that redirects the call to an appropriate destination based on call associated information and interaction with a caller.

Features such as *do not disturb*, *forwarding*, *call offering*, and *auto-answer* provide mechanisms for the automated handling of attempts to connect calls to specific devices.

Calls may be *parked* at specific devices and retrieved using services such as *directed pick up call* and *group pickup call*. Calls may be suspended, or placed in the *hold* state, using the *hold call* service or the *consultation call* service. The *consultation call* service creates a new call after suspending the first call. The *alternate call* service can be used to toggle between the calls, and the *reconnect call* service can be used to drop one call and return to the one that is on hold. Depending on the *consult purpose* specified (if necessary for a given telephone system), the first call can be transferred to the second called device using the *transfer call* service, or all three devices can be merged onto the same call using the *conference call* service. The *single step transfer call* and *single step conference call* services provide single-step versions of these functions. If *silent participation* is supported, devices can be added to calls in such a way that they can listen but not participate in a call.

If call processing is unable to connect a call and its routing options are exhausted, the call may either be *recalled* to the last device that routed it, or it may be stalled. If call processing for a call is stalled, the call is said to have *failed*. The *camp on call*, *call back call-related*, *call back message call-related*, and *intrude call* services are all *call completion services* for handling a failed call.

Finally, when nothing further is required of a connection or an entire call, it can be cleared using the *clear connection* or *clear call* service, respectively.

5.
Telephony Equipment and Network Services

We turn now to looking at how actual telephony products and services are implemented and how they reflect the abstraction we explored in Chapters 3 and 4. The concrete examples provided in this chapter breathe more life into the abstractions presented earlier.

Each individual telephony product is a telephone system, that is, a set of telephony resources. Each telephone system generally is a subsystem of a larger telephone system, and virtually all telephone systems are ultimately connected to, and thus are part of, the world-wide telephone network (the PSTN).

5.1. Telephony Products

Telephony products can be categorized generally as being either telephone switches or telephone station equipment, or as being an add-on or peripheral to one or the other.

5.1.1. Telephone Switches and Switch Peripherals

 A *telephone switch*, or just *switch* for short, is a telephony product characterized as an implementation of telephony resources that, at a minimum, includes call processing and switching resources, and at least two station or network interface type devices.

A generic telephone switch is illustrated in Figure 5-1.

Figure 5-1. Generic telephone switch

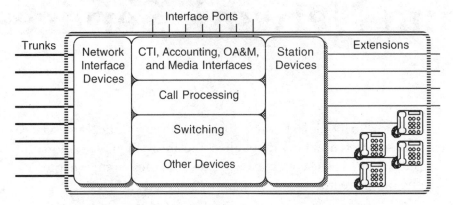

A switch may consist of any number and combination of telephony resources built around the core functionality of call processing and switching resources (which may be minimal in some cases).

 The switch makes connections to other telephony products through *lines* that represent the cabling, wireless transceivers, or other means of connection. Lines fall into two categories, *extension* (or subscriber) lines and *trunks*.

The point where a line is connected to a switch is often referred to as a *port*. In addition to the connections a switch can make using lines, the switch may connect with other special peripherals through *interface ports*.

Extension or Subscriber Lines

Extensions are lines that correspond to a switch's station devices. From the perspective of the switch, each extension line corresponds to the logical element of one station device. When the switch is part of the public network, these lines are referred to as *subscriber lines*, *subscriber loops*, or *CO lines*.

If a given switch is able to detect and control one or more physical telephones attached to a particular line, the corresponding station device either includes a physical element, or is associated with other devices containing physical elements through a device configuration. In this case, the switch keeps track both of the station device with the single logical element corresponding to the line, and of all the physical element station devices associated with the port.

Additional telephony resources making up an operational switch—including any CTI, OA&M, accounting, and media interfaces—may be either built-in or implemented as add-on peripherals. Depending on the implementation, peripherals may be attached through the switch's internal connections, or may be attached using line ports. In these cases, the switch keeps track of the other device types and resources associated with the port.

Trunks

Trunks are lines that correspond to network interface devices. They are used to connect to one or more external networks by attaching to switches that are part of the external network.

A trunk associated with one switch can be connected to either a trunk or a subscriber line of another switch. In other words, the network interface device of the first switch may correspond either to a station device or a network interface device of the second switch.

Interface Ports

Interface ports are the means for physically connecting computers or other peripherals to a switch's CTI interface, OA&M interface, accounting interface, or media interfaces (in switch implementations that do not use line ports for this purpose).

5.1.2. Telephone Station Equipment

Telephone station equipment, including telephone station peripherals, are telephony products that are connected to the extension/subscriber lines of a switch. They are fundamentally implementations of station devices, along with associated switching, call processing, media service, and interface (CTI, OA&M, accounting, and media) resources. A particular piece of station equipment may be modeled one of three ways, depending upon the context:

1. If it is modeled independently of any connection to a telephone system, it is viewed simply as a physical device element made up of the appropriate physical element components.

2. If it is connected to a switch, the telephone station may be modeled from the switch's perspective as a physical element associated with one or more of the logical station devices in the switch (as described in section 5.1.1 above). This view of a telephone station is illustrated in Figure 5-2.

Figure 5-2. Generic telephone station

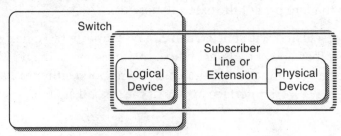

3. If it is receiving telephone service from a switch but is viewed as being a standalone telephone system, the station equipment is modeled as a complete set of telephony resources. This includes a station device, or device configuration, that corresponds to the station's physical element and associated logical device element(s), along with associated telephony resources that include switching, call processing, media services, and any CTI, OA&M, and media interfaces. In most cases, call processing in the telephone station accepts commands from the physical element interface and/or the CTI interface, and translates them into appropriate commands for the switch using the telephone station's network interface device as a proxy. This view of a telephone station is illustrated in Figure 5-3.

Figure 5-3. Generic telephone station modeled from station's perspective

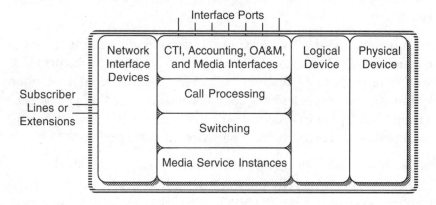

Refer to Chapter 3, section 3.6 for a complete description of physical element components.

5.1.3. Customer Premises Equipment (CPE) Options

Customer premises equipment, or *CPE* for short, refers to the telephony products that you purchase or lease in order to assemble a private telephone system of any size.

The trunks or lines that you already have and those that are available to you from various telephone companies (carriers) will, in part, dictate what types of telephony products are applicable to your needs. Depending on the requirements, your telephone system might be as small as a single telephone to be connected to a subscriber line from your local telephone company or cellular carrier, or it may be as complex as a global private network.

5.2. Telephone Company Services

If you are assembling a new telephone system for your home or business, or are assessing your existing system, one of the first steps is to determine what services you want to obtain from your local exchange carrier(s), that is, your local telephone company or companies, and any other carriers with which you wish to have connections.

The functionality of your telephone system (whether it is just a single telephone or a collection of switches forming a private network) will be a combination of the features and services built into your telephony equipment, and any telephony features and services that you *subscribe to* from the telephone company (or companies) that your equipment can access using corresponding network interface devices.

As described earlier (in Chapter 3, section 3.4) a LEC, or local exchange carrier, generally provides the connection(s) between your telephone system and the PSTN (the network external to your system). In other words, the LEC owns the switch or switches that extend lines or trunks to your telephone equipment, for local calling and typically for access to IXCs (interexchange or long-distance carriers). You may, however, connect directly to one or more IXCs through what is referred to as carrier bypass. (Note that the notion of bypass applies only in the context of competing carriers. In countries where a single telephone company provides all of these services, this notion does not apply.)

A switch owned by the LEC to which your lines or trunks connect is known as a *central office switch*, (*CO switch* for short), or a *public exchange*. The combination of line interfaces and telephony features and services available for subscription on a given CO switch represents the *service offerings*.

5.2.1. Individual Subscriber Lines

One way to obtain telephone service from a carrier is simply to order one or more *individual subscriber lines*. These are effectively extensions associated with the carrier's CO switch, and each has a unique telephone number. At a minimum, these lines generally offer the functionality of placing, answering, and dropping calls (POTS functionality).

Your telephone company also may offer various supplementary telephony features and services that can be activated for some fee. The set of available services, the names under which they are marketed, and their cost vary widely between countries, states, telephone companies, and between different brands of CO switches used by a carrier. A well-defined group of functions in North America are referred to as *CLASS*, or *Custom Local Area Signaling Services*. These services include callerID and features that take advantage of other call associated information.

5.2.2. Centrex Services

Centrex services are a variation on individual subscriber lines, wherein an entire group of lines is purchased at a time and the lines all behave as if they were connected to a private (CPE) switch. Centrex generally is positioned as an alternative to buying your own telephone system; a portion of the central office switch is dedicated to your telephones and simulates the operation of a self-contained switch.

The chief advantage and disadvantage of Centrex are opposite sides of the same coin. The advantage is that you are able to rely on the telephone company to manage and administer your telephone system for you. The disadvantage is that you must rely on them.

Organizations most likely to use Centrex are those that have multiple physical locations in the same geographical area, all of which should be connected to the same telephone system. If all of the extensions in all of the sites would have to be connected via a CO switch in any case, Centrex generally is a more efficient approach than having a CPE switch. Theoretically the overall capacity of the CO switch is likely to mean fewer limitations to the expansion of your telephone system than a smaller CPE switch, but this is not always the case.

5.2.3. Combination Trunks

A *combination trunk* is a trunk line that allows *two-way* (inbound and outbound) operation. In other words, it is basically just an individual subscriber line that corresponds to the network interface device of your switch, rather than the line associated with a piece of telephone station equipment.

Combination trunks can be formed into an *incoming service group* (ISG), which is a hunt group within the CO switch that allows an incoming call to use the next available trunk (network interface device) on the receiving switch. Your business's main number is assigned to the hunt group so that your callers will never get a busy tone, assuming you have enough trunks in the ISG to handle your peak number of simultaneous incoming calls.

5.2.4. DID Trunks

DID trunks are trunks that support only incoming calls and that support the DID feature described in Chapter 4, section 4.6.2. DID allows an incoming call to be directed automatically to the appropriate extension by having the calling switch indicate the desired extension to the answering switch; it does so by dialing the last two, three, or four digits of the number originally dialed. DID trunks are always provided as a set of lines associated with an ISG (incoming service group or hunt group) that directs each new incoming DID call to the next available DID trunk.

5.2.5. Tie Lines and Private Networks

One type of service you may obtain from a carrier is a *tie line* that connects two CPE switches in different locations, thereby creating an extended private network. This type of point-to-point line is treated as a trunk by both of the CPE switches concerned, and it is up to you to set up the dial plan rules and network interface groups for each switch to determine how the trunks will be used.

For example, if a switch in Toronto and a switch in Ottawa are linked with a tie line, extensions on the switch in Ottawa can place calls to extensions on the switch in Toronto as if they were themselves in Toronto. This is illustrated in Figure 5-4.

Figure 5-4. Tie line example

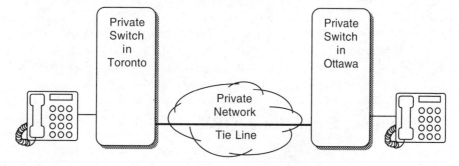

A tie line is referred to as a *dedicated circuit* or *leased line* if it is a facility that is for the exclusive use of the subscriber, so that it is guaranteed to be available at all times even though it is passing through one or more carriers' networks. A virtual private network (see Chapter 3, section 3.4) provides tie lines without physically dedicating a particular set of facilities. Instead, the facilities are allocated on an as-needed basis.

5.2.6. Foreign Exchange (FX) Lines

A *foreign exchange line* is like a tie line except that it connects your telephone system directly to a CO switch (public exchange) in a distant location. A foreign exchange line allows your telephone system to receive and place calls locally in two or more different geographical locations. This is typically implemented for trunks, but it could be applied to a subscriber line.

For example, if a switch in San Jose has a foreign exchange line to Sacramento, callers in Sacramento can place incoming calls to the switch using the local number corresponding to the switch. Likewise, extensions on the switch can place calls to Sacramento local numbers without incurring toll charges. This is illustrated in Figure 5-5.

Figure 5-5. FX line example

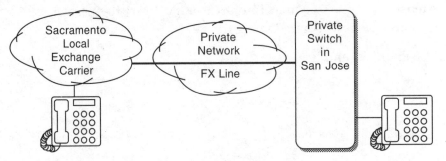

5.2.7. Off-Premises Extensions (OPX)

An *off-premises extension* is a dedicated circuit that connects an extension on a CPE switch to a subscriber line associated with a particular CO switch. This is illustrated in Figure 5-6. A telephone station that is connected in this fashion behaves in every way as though it were directly connected to the switch on which it is an extension.

One example of OPX use would be the arrangement used by a doctor associated with a small clinic, and who also manages a practice at home. In this example, one extension on the clinic's switch has its telephone station at the doctor's home practice.

Another popular use of the OPX feature involves telecommuting. A car rental company could allow its reservations agents to work from home by placing extensions corresponding to devices in the call

Figure 5-6. OPX line example

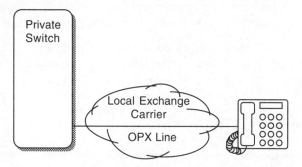

center's ACD group at their homes. The reservation agents then could log onto the ACD group, handle customer calls for reservations requests, and log off at the end of their shifts—all from home.

5.2.8. Toll-Free Numbers

Toll-free numbers allow a caller anywhere within a defined geographical area to dial the number without paying long-distance charges. The resulting call is then routed to any designated line or ISG (incoming service group or hunt group) and is billed to the called number.

In North America, toll-free numbers are referred to as *IN-WATS* (INward Wide Area Telephone Service). Numbers are dialed using 1-800 or 1-888 followed by the unique seven digits of the toll-free number.

These numbers can be configured to deliver calls to lines dedicated to handling toll-free numbers, or can be directed to any general-purpose line that supports incoming calls. In addition, subscribers to these services also have the option to obtain the DNIS (dialed number identification service) and/or ANI (automatic number identification) information with each toll-free call placed, independent of any callerID blocking in place on the caller's logical device. The mechanisms used for delivering this call associated information vary, depending on how call delivery is configured and the telephone company providing the service. (See Chapter 4, sections 4.5.1 and 4.5.2 for more information on ANI and DNIS.)

In many ways, the reliable delivery of ANI and DNIS information is the most valuable part of toll-free service. The expense of these lines may not be justifiable purely on the basis of customer/caller convenience; the cost savings that result from installing a CTI system that takes advantage of this information, however, often can justify the cost of obtaining an 800 or 888 number.

5.2.9. Alternate Wireline Providers

All of the services described in this section are provided by traditional telephony common carriers. These are companies and organizations that manage the switches making up the PSTN (the worldwide public telephone network). In some locations, however, these services or equivalent services also can be accessed through *alternate wireline providers*. These are organizations that provide an alternative communications path for delivering telephony services. Cable TV companies are among the first to begin providing telephony services; power and other utility companies may do likewise. Internet service providers and those providing telephony services through the Internet also represent another form of alternate telephone company.

5.2.10. Alternate Non-wireline Providers

A very popular alternative to the traditional delivery of telephone services through copper or fiber cable is wireless telephone service. Cellular carriers are the most prevalent form of wireless providers, but other forms of wireless subscriber loop are also common.

5.3. Line Interfaces

Switches, telephones, and line-based peripherals connect with one another using connections referred to as lines and trunks, or alternatively as *communications facilities* or *transmission facilities*. Subscriber lines also are referred to as *subscriber loops* or *local loops*. In this section we'll explore the various forms that these facilities can take.

5.3.1. Telephony Media and Signaling

Lines carry all of the information necessary to convey control information, call associated information, and media stream(s) associated with a call. The *capacity* of a particular line can be measured in terms of the number of media stream channels and other *signaling channels* it has the capacity to carry. Another measure is the overall *bandwidth*[5-1] of the facility.

The principal payload of a line is the one or more media streams that it carries. In addition, command and control information including the following must be transmitted:

- Signals to reserve the use of a media stream channel and to indicate the presence of a new call on a media stream channel;

- Commands from telephone station equipment to execute services; and

- Physical element control and status information for each connected telephone station on lines that allow a switch to detect and control telephone station equipment.

Finally, call associated information, including the following, may need to be conveyed for calls being carried by a given line.

- Direct Inward Dialing (DID)

- CallerID

- ANI

- DNIS

- Correlator data

- User data

5-1. Bandwidth — Technically the term *bandwidth* refers to the range of frequencies a facility can carry. In common usage, it is a measure of the information-carrying capacity in thousands of bits per second (kbps) or millions of bits per second (Mbps).

These different types of control and call associated information may be communicated in-band or out-of-band.

In-band Signaling

In-band signaling refers to a technique in which commands and information share the media stream channel for a particular call. The information is encoded as breaks in the channel, special tones or tone sequences, or modulated data. A significant disadvantage of in-band signaling is that the contents of the actual media stream can be mistaken for signaling information.[5-2]

Out-of-band Signaling

Out-of-band signaling refers to techniques where separate electrical circuits or separate signaling channels are used to keep control and status information separate from media stream information. Separate signaling channels may be established over the same circuit as the media stream channel through multiplexing (see sidebar "Multiplexing").

Quality of Service (QoS)

The media stream channels supported by a particular transmission facility may be implemented using analog or digital technology. Analog media stream channels support only voice calls (3.1 kHz of bandwidth for speech and modulated data). Transmission facilities supporting digital media stream channels allow both voice calls and digital data calls. Voice calls are encoded to pass through the digital

---

5-2. In-band signaling — The best illustration of the negative consequences of in-band signaling is the so-called Captain Crunch™ whistle. A novelty whistle included in a certain brand of breakfast cereal was found to have a pitch of exactly 2600 Hz. At the time, this was precisely one of the frequencies that AT&T had chosen to use as part of its long-distance network's in-band signaling protocol. People could connect to AT&T's long-distance network, blow the whistle, and then dial free calls using tone dialing (assuming they knew the rest of the protocol). Since then, the vast majority of the world's long-distance network infrastructure has switched to out-of-band signaling.

network in a fashion such that the voice can be reconstituted in analog form as needed. In this case, an indication as to whether a particular call is a voice call or a digital data call becomes yet another piece of call associated information.

Multiplexing

Multiplexing is a technique for dividing the bandwidth available on a single facility into multiple channels.

One approach to multiplexing is referred to as *frequency domain multiplexing* (*FDM*). In this approach, each channel is assigned a frequency band and its data is encoded within that band. (This is where the term *bandwidth* as description of channel size was derived.) For example, FDM is the technique used for multiplexing multiple TV channels on a single coaxial cable.

The multiplexing approach commonly used in telephony is *time division multiplexing* (*TDM*), in which the bandwidth of the channel is divided into individual time slots and each channel uses a share of the time slots. Each channel being multiplexed takes its turn to insert a few bits of information into its slots at the appropriate interval, as shown below.

For example, if six channels of equal size are being multiplexed together, the 3rd, 9th, 15th, 21st, etc., time slots would be used by the third channel.

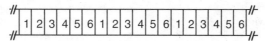

If the channels are of different capacities, some channels have more time slots than others. For example, if three channels are to be multiplexed together but the first two are both four times the size of the third, then the pattern for channel allocation would be such that of every 18 time slots, channels 1 and 2 would each have eight slots and channel 3 would have two.

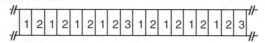

5.3.2. Analog Lines

Analog lines are the oldest form of subscriber loop technology. In fact, a whole family of different analog line implementations exists, each one varying in terms of its support for different types of associated information and the way that this information is conveyed.

The one element that all analog lines have in common is that a single pair of wires[5-3] forming an electrical circuit carries only one media stream channel. Figure 5-7 shows an analog line or *local loop*. Analog telephone equipment connects to the analog loop through a hookswitch. When the hookswitch is open (as shown) the telephone is said to be *on-hook* (in a simple mechanical phone, the receiver is in its cradle), and the electrical circuit is open so no current flows. When the hookswitch is closed (the telephone is said to be *off-hook*), electric current flows through the resulting completed circuit. The media stream from the switch and from the microphone of the telephone is then applied to the line. Both the telephone and the switch are able to listen to the media stream on the analog loop, using hybrid circuits that differentiate between the portions of the media stream that are being generated and received at the same point on the loop.

Figure 5-7. Analog local loop

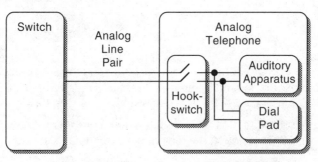

5-3. Tip and Ring — For historical reasons having to do with the tip and collar (ring) of the plug used in manual patch-panel switchboards, the two wires used are arbitrarily referred to as *tip* and *ring*, or "T" and "R." The term *twisted pair* refers to the fact that the pair of wires forming the circuit is twisted together to minimize electrical interference and crosstalk.

Figure 5-8 shows the same analog loop with multiple telephone stations attached or *physically bridged* onto the line. One of the chief advantages of analog loop technology is that it offers a reasonably high degree of flexibility with respect to the way that telephone sets can be added or removed. The loop can be extended and added to, up to the limits of the switch's signal strength.

Figure 5-8. Analog local loop with bridged telephone stations

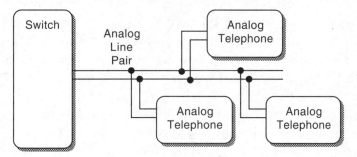

Loop Start Signaling

When all the telephone stations on a given loop are on-hook, no loop current is flowing and there is no media stream, and therefore there is no means of conveying in-band signaling. As a result, analog lines require additional signaling mechanisms for use while in this state.

Loop start is the most common technique used to allow a telephone station to request service from the switch on an analog local loop. With loop start signaling, the telephone set goes off-hook, which closes the electrical circuit and allows current to flow. The switch detects this flow of loop current, creates a new call with a connection in the *initiated* state, and applies dial tone to the media stream.

On the other hand, the switch indicates that it is trying to connect a call in the *ringing* mode of the *alerting* state by generating an alternating current (AC) on a local loop that is open (i.e., no telephones are off-hook, so the circuit is open). The AC voltage causes the ringers in all the telephones on the loop to signal that an attempt is being made to connect a call. When one of the telephones goes off-hook, the resulting flow of loop current indicates that the *answer call* service is to be invoked. The switch than transitions the connection from the *alerting* state to the *connected* state (stopping the AC ringing signal) and delivers the media stream(s) for the call using the media stream channel represented by the analog line. The switch assumes that it can begin using the line at any time.

Ground Start Signaling

Ground start is another technique used for requesting service on analog loops. Ground start lines are typically used for two-way combination trunks. Two-way operation requires that use of the line be controlled in such a way that both switches do not try to use the same trunk at the same time.[5-4] In ground start signaling, each switch is assigned one of the two wires; each indicates that it wants to reserve the line for use by grounding its assigned wire (which is detected as the flow of electrical current in one of the wires, but not around the loop). The grounding action allows one switch to signal the other that it wants to use the line so that the second one doesn't. In every other respect, ground start lines work the same way that loop start lines work.

Wink Start Signaling

Wink start is the third common technique for performing signaling on an analog line. Wink start is used primarily for DID trunks that are used only for incoming calls from a CO switch to a CPE switch, and thus don't involve supplying dial tone or entering the *initiated* state. Wink start is quite different from the other configurations of analog lines because it is the CO switch that uses the hookswitch and it is the CPE switch that detects the flow of loop current used by the CO switch to indicate the presence of an incoming call. When the CPE switch detects the loop current, it *winks* by reversing the polarity of the voltage on the loop. This signals the CO switch to provide the DID information associated with the call. When the CPE switch has received the DID digits, it winks again to indicate that it is ready to accept the actual media stream associated with the call.

5-4. Glare — The situation where two switches inadvertently seize the same line for two different calls is called *glare*. *Ground start loops* prevent glare.

POTS

POTS, or *plain old telephone service*, refers to the lowest common denominator of analog telephone services. Telephony features on a POTS line are restricted to making, answering, and dropping calls.

To make calls, either the loop start or ground start technique is used, depending on the type of line. This request for service corresponds to invoking the *make call* service with no digits. The digits to be dialed are conveyed on the analog local loop using either pulses or DTMF (touchtones). Pulse dialing involves breaking the loop current approximately ten times per second to count out each digit desired, with a pause of between 0.6 and 0.9 seconds between the digits. DTMF dialing simply generates the appropriate tone for each of the desired digits. All dialing on a POTS line is multi-stage, and each digit conveyed is a request for the *dial digits* service.

The telephone station conveys the command to clear the connection by putting the telephone station on-hook. When every telephone station on the local loop is on-hook, the electrical circuit is broken and the switch interprets this as a request for the *clear connection* service for the logical device in question.

In general, the behavior of a logical device element representing a POTS line corresponds to the behavior described for interdependent–shared–bridged appearances (in Chapter 3, section 3.8.3). If a given POTS line is dedicated to just one telephone station, however, and there is no physical bridging taking place, then the logical device may be modeled as having non-addressable standard appearances.

DTMF Feature Codes

One enhancement to POTS that is supported by most switches involves the ability to access many, if not all, of the telephony features and services it supports by issuing commands using special sequences of DTMF digits.

There is no universal standard for the assignment of these codes to telephony features and services, so each telephone company and switch vendor assigns them as needed. These sequences usually begin with the digit "*" or "#" in order to distinguish them from digit sequences to dial. This feature works by having the switch look at the first digit pressed when a new call is connected to the device in the *initiated* state. If the first digit is "*" (or "#" depending on the switch implementation), the rest of the sequence is interpreted as a command. Otherwise all tones are interpreted as digits to be dialed.

For North America, Bellcore has defined some feature codes for use by all public carriers. These include the sequence "*67" which invokes the *set callerID* service to deactivate sending of callerID information with the next call and "*82" which uses the *set caller ID* service to activate sending of callerID information with the next call.

Hookswitch Flash

Nearly all switches support the use of *flash hook*, or just *flash* for short, on their analog lines. This involves detecting short breaks in loop current during an active call as an invocation of the *consultation call* service if there are no other connections at the device, or the *conference call* or *alternate call* services if there is another connection at the device that is *alerting* or on *hold*.

If there are no other connections at the device, the flash is interpreted as a request for the *consultation call* service with no digits dialed (just as with POTS *make call*). The active connection is placed in the *hold* state and dial tone is heard on the media stream for the new call (which is associated with the device using a connection in the *initiated* state). If the switch support DTMF feature codes, then the first digits are screened to determine if a command is being issued or if a dial string is being entered. Depending on the digits pressed, a feature or service is invoked or a new call is originated.

If there is another connection at the logical device corresponding to the analog line when the hookswitch is flashed, the flash is interpreted as invoking either the *alternate call* service or the *conference call* service, depending on the state of the other connection, the switch implementation, and the class of service assigned to the device. Typically if the second connection is in the *alerting* state, the *alternate call* service is used, which places the first call on *hold* and transitions the second call to the *connected* state. If the second connection is in the *hold* state, the *conference call* service typically is the one invoked. This merges the two calls into a single call with everyone participating.

Most systems also support the *transfer call* and *single step transfer call* services through a related mechanism. If flash was used to place a connection on *hold* and to originate a new connection, dropping the call before the new call is answered is treated as a request for the *single step conference call* service. Dropping the call after it is answered is treated as a *transfer call* request.

CallerID

One of the most important pieces of call associated information for CTI solutions is callerID. Despite the limitations of analog lines, callerID information can be delivered and most telephone companies now include callerID among their service offerings for analog lines. CallerID information is sent by the CO switch as a burst of modulated data between the first and second ring cycles as a call is being presented. To use callerID, telephone station equipment must be equipped with appropriate electronics for intercepting and decoding the callerID information, and the phone must not be answered until after the second ring has started. (See Chapter 4, section 4.5.1 for more information on ANI and callerID.)

Distinctive Ringing

Some analog lines are able to indicate the original dialed number associated with a call that is being presented to a particular device using the *distinctive ringing* feature. This feature is effectively an analog line version of DNIS that involves using a different ringer pattern (the cadence of the AC signals making up a ring cycle) to correspond to each of the subscribed-to numbers that are associated with the analog line in question. The mapping between numbers and ringer patterns is specific to each and every line, so this feature is rarely taken advantage of in CTI systems. To do so requires special hardware capable of detecting different ring patterns and prior knowledge of the assignments of ringer patterns to numbers. (See Chapter 4, section 4.5.2 and section 5.2.8 above for information on DNIS.)

Call Waiting Indication

Call waiting indication is used in conjunction with the flash hook capability. It involves allowing for a second call to be presented to the logical device corresponding to an analog line while a first call is active. The presence of the second call in the *alerting* state is signaled using a special in-band signaling tone that is inserted into the media stream channel on the analog line. The second call can be answered by issuing the *alternate call* service using a hookflash, or can be ignored.

The latest variation on call waiting involves delivering callerID information for the second call using additional in-band signaling. While having this information is very useful, the disruption to the active call, resulting from the in-band signaling that otherwise blocks the active media stream, can be significant.[5-5]

Proprietary Second Pair Signaling

Despite the fact that analog local loops (in fact, most local loops, period) require only a single pair of wires, standard telephone cable has at least two pairs of conductors, and the standard North American telephone jack, the RJ-11, provides a connection for both pairs. Certain switches put the normally unused pair of wires to work as a separate, independent signaling channel.

5-5. Call waiting and in-band signaling — Call waiting illustrates a second disadvantage of in-band signaling. If the switch assumes that a human is listening and can take action, it may generate in-band tones at any time. This can disrupt modulated data connections (i.e., modem or fax transmissions) that might be present on the line. For this reason, most systems that support call waiting also support DTMF commands to disable the feature.

These implementations support full POTS compatibility with the standard pair of wires so that any existing analog station equipment will work. Special proprietary phones compatible with the proprietary protocol on the second pair, however, are specially handled by the switch when detected. These products use the second independent (typically digital) channel in order to allow full functionality on the analog loop. Information conveyed includes all available call information, status and control information for the physical device, etc. This arrangement is illustrated in Figure 5-9.

Figure 5-9. Proprietary second pair

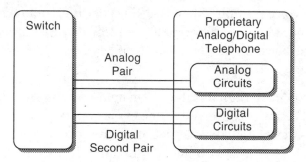

Analog Telephone Station Equipment

Analog telephone station equipment is anything that can be connected to an analog telephone line. This includes:

- Analog telephone sets

- Fax machines

- Data modems

- Fax modems

- Low-speed video phone telephones

Individual devices typically combine multiple functions and can be found in a limitless variety of form factors.

The portions of these products that correspond to the components of a telephone set represent each item's physical element. Other portions (such as a fax machine's scanner, printer, and modem) are considered media services associated with the device. Telephone stations are discussed in more detail in section 5.7.

5.3.3. ISDN-BRI

Integrated services digital network, or *ISDN*, refers to the standards defined for providing subscriber access to the full capabilities of the digital portion of the telephone network while still interoperating with the analog/voice portion. There is a series of different line interface standards that represent different levels of ISDN capability and capacity. The basic version is called the ISDN *Basic Rate Interface*, or *BRI* for short.

Basic rate ISDN is a digital (rather than analog) subscriber loop technology. BRI was conceived as the evolutionary path for analog lines. It was designed to be compatible with most existing analog subscriber loop wiring, with the intention that ISDN basic rate service would be a simple upgrade for people using analog lines and wanting to move to digital technology.

With ISDN, all the signaling, control, and media stream information and other call associated information is sent in digital form. BRI uses multiplexing to provide three channels:

- One *signaling* channel referred to as the "D" channel. This channel is a 16 kbps (bidirectional) channel; and

- Two *bearer* channels, referred to as "B" channels. Each B channel is a 64 kbps (bidirectional) channel.

An ISDN BRI line can be visualized as shown in Figure 5-10.

The B channels are media stream channels. The D channel is used for all other information exchanged between telephone station equipment and the switch. This means that all signaling is out-of-band. The protocols used for information exchanged on the D channel are

Figure 5-10. ISDN BRI

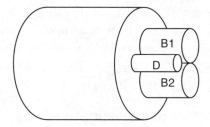

defined as part of the international ISDN standard. The D channel is capable of conveying all of the call associated information[5-6] and requests for most of the popular telephony features and services. In fact, in most locations callerID service was available on ISDN BRI long before it was available on analog lines.

Both bearer channels are capable of carrying both voice calls (including modulated data such as modem and fax) and digital data calls.[5-7] The combination of uses subscribed to for a specific channel is referred to as the channel's *bearer capabilities*. Depending on how the switch and line are configured for use, restrictions may be applied to the use of each B channel for voice or data calls. This is often done, for example, to prevent both channels from being used for data, leaving no channel for a voice call if one had to be made. The bearer channels are typically used independently so that, for example, one may be used for a voice call while another is being used to connect to the Internet using a digital data call. It is also possible to use both B channels for a single digital data call (128 kbps) using a standard known as *BONDING*, which in reality still establishes two independent calls to another ISDN interface, but uses protocols to synchronize the contents of both calls to appear as a single digital data media stream in each direction.

5-6. Versit UUIE — The ISDN signaling protocol defines a mechanism called User to User Information Elements (UUIE) that allow the exchange of user data. Versit has defined a means of encoding both correlator data and user data into the UUIE. This is referred to as *Versit UUIE* and is documented in Appendix A of Volume 3 of the *Versit CTI Encyclopedia*.

5-7. Packet switching — Both bearer and signaling channels can also be used for carrying X.25 packet-switched data.

Installation and Configuration

Technically speaking, ISDN BRI actually refers to a whole family of individual line interfaces that work together. Attaching telephone stations (or *terminal equipment* as it is known in ISDN) to a BRI line is not as easy as attaching equipment to an analog line. Figure 5-11 shows a generic ISDN BRI installation.

Figure 5-11. ISDN BRI interface points and equipment

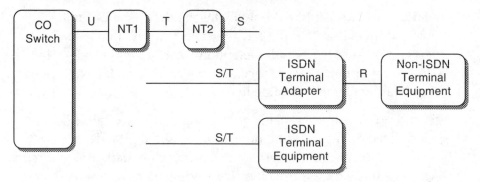

The BRI line directly from a central office switch is known as the "U" interface. The U interface is capable of encoding[5-8] all of the ISDN BRI digital information on a single pair of wires that is compatible with most of the analog subscriber loops currently in use.

The ISDN standard also defines an "S" and a "T" interface. These are actually the same interface electrically, but the letter indicates what type of equipment is exposing the interface. The ISDN standard defines a piece of equipment that converts the U interface to the T interface as a *Network Termination–Type 1*, or *NT1*. If the line is to be

---

5-8. 2B1Q — The encoding used for the U interface is not internationally defined. Each country defines the U interface to be used on subscriber lines provided by local exchange carriers. In North America the standard for the U interface is referred to as 2B1Q (ANSI standard T1.206); this standard was not established, however, until after vendors had already begun shipping switches that used a different U interface encoding referred to as AMI. While AMI is now obsolete, older equipment and lines may still be in use.

connected to a switch that in turn has ISDN basic rate lines, those lines are considered to be S lines and the switch, for purposes of the ISDN standard, is considered a *Network Termination –Type 2*, or *NT2*.

Terminal equipment is attached to the S/T interface. (S and T are equivalent from the perspective of terminal equipment.) The S/T interface is a four-wire interface (two pairs of wires) and is the interface provided on generic ISDN telephone station devices. A big advantage of ISDN is that terminal equipment can be used anywhere in the world because this interface is standardized internationally.

A *terminal adapter*, or *TA* for short, is a piece of ISDN terminal equipment that allows one or more pieces of non-ISDN equipment to be adapted to the ISDN line. The number of adaptations possible is unbounded and thus is not standardized. The "R" interface generically represent all of these non-ISDN connections.

Power

One important difference between ISDN and analog lines is that analog lines utilize loop current from the switch, so that all the power needed by the telephone station can be obtained from the line itself.

With ISDN lines, the line (U interface) only carries low-voltage signaling. This means that all power for operating the various devices that allow for a functional ISDN line (NT1, etc.) must be powered independently. If there is a power failure and the NT1, NT2, and other ISDN station equipment lose power, the ISDN line cannot be used. For this reason it is generally recommended that, unless a given ISDN line is intended only as a spare line or for recreational purposes, associated equipment should be powered through an *uninteruptable power supply* (*UPS*).

Multi-Point Installations

The S/T interface supports the ability for up to eight pieces of terminal equipment to connect to the same S/T interface, in much the same way that multiple analog telephones can be connected to the same analog line. This is shown in Figure 5-12.

Figure 5-12. ISDN multi-point

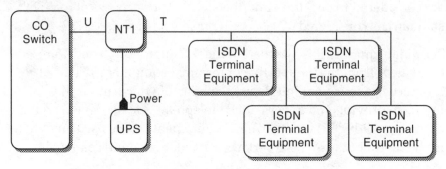

ISDN Telephone Station Equipment

ISDN telephone sets connect directly to the S/T interface. They take direct advantage of the D channel for receiving notification of connection attempts, placing calls, requesting telephony features and services, and receiving call associated information such as callerID. They can access either B channel for voice media stream data (assuming that the line is configured to use both B channels for voice), but can only listen and talk on one channel at any instant.

While other types of standalone ISDN station equipment do exist, such as videoconferencing devices, most ISDN station equipment is in the form of terminal adapters that integrate with other communications products.

Terminal Adapters

Terminal adapters, or *TAs* for short, typically are standalone peripherals or add-in cards for personal computers. They also may appear in the form of cards inside other forms of communications equipment or inside an ISDN telephone set.

Terminal adapters intended for use with personal computers (either as add-in cards or as external peripherals) typically are used to establish digital data calls for connecting to the Internet, exchanging large documents, and doing videoconferencing.

Some terminal adapters can handle both voice and digital data calls by using built-in modems[5-9] so that they can also exchange faxes and data with analog fax and data modem equipment in the voice telephone network.

Analog Telephone Station Equipment

A very popular R interface for a terminal adapter is an analog telephone line circuit. Terminal adapters may include one or two analog line interfaces that correspond to one or either B channel. This allows existing analog telephone station equipment (telephones, fax machines, etc.) to be attached to an ISDN line. This is shown in Figure 5-13.

Figure 5-13. Analog stations on an ISDN line

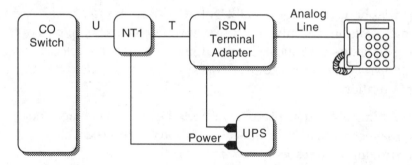

Built-in NT1s

Many ISDN terminal adapters and other station equipment incorporate the NT1, so that they plug directly into the U interface. This is done to avoid the expense and complexity of having two separate devices. There is a trade-off to be made, unfortunately, because only one such device can be attached to a given U interface, so

5-9. ISDN modems — Terminal adapters that have built-in modems are generally called *ISDN modems*. People sometimes misuse this term to refer to terminal adapters that have the external appearance of traditional analog modem peripherals but don't actually have modem functionality. This is not correct and can be confusing.

one can have, at most, one of these types of devices per line. In addition, unless the device also supports a T interface (from the built-in NT1), other S/T equipment cannot be used and multi-point operation is not supported.

Modeling ISDN

There are many different ways that ISDN can be modeled. Different switch implementors have pursued different approaches.

One approach is to model the ISDN line as a single logical device and to model each ISDN bearer channel as an independent media stream channel that is available for use by an appearance. Another approach is to model each ISDN bearer channel as a distinct logical device. In either case, appearances may be addressable or non-addressable.

The telephone stations on an ISDN line correspond to physical device elements. If there is only one, it may be modeled as being part of the same device as the logical element. Otherwise, it is modeled as part of a *multiple logical elements* device configuration or a *bridged* device configuration.

Sub-addressing can be used, if supported by the switch and the external network, to indicate which station (i.e., physical element) is to hear ringing with respect to an *alerting* call.

5.3.4. Proprietary Digital Subscriber Loops

Proprietary digital loops are digital lines from switches that are proprietary to a particular switch vendor. There too many proprietary implementations to enumerate, as each switch vendor typically has one or many that they have invented. Proprietary telephone station equipment is compatible only with the specific flavor of proprietary digital subscriber loop technology for which it was built.

Despite all of these incompatible designs, the underlying concepts are usually all very comparable to ISDN, if not actually based on ISDN. Typically, most proprietary digital loops have a signaling channel and one, two, or more bearer channels (although the additional bearer channels usually are either never used or used only for digital data). This is illustrated in Figure 5-14.

Figure 5-14. Proprietary digital loops

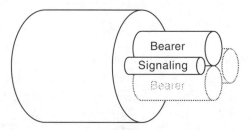

The key differences between implementations involve the range of functionality that can be accessed and controlled (in either direction) by the signaling channel and the bandwidth associated with the bearer channels. (If supported, this is usually 56 kbps or 64 kbps in order to take advantage of standard voice media stream encodings.) For example, one vendor's signaling protocol may provide complete functionality for control and observation over the components of the physical telephone set while another's may support full access to every imaginable telephony feature and service.

5.3.5. T-1, E-1, and Other Multiplexed Spans

The facility for carrying a channel from one switch to another is called a trunk, and in most cases there is a desire to maximize the number of trunks that can be delivered while using as few physical wires as possible. The way this generally is accomplished is by using a high-speed digital connection, called a *span*, between two points and using time division multiplexing (see sidebar "Multiplexing" on page 261) to divide it into multiple fixed-bandwidth channels.

There are many variations of digital trunking technology available, depending on the capacity desired. The internationally agreed-upon channel size is 64 kbps and a channel of this size is referred to as a *DS-0* channel (for digital-signal level 0). DS-0 channels are then multiplexed together to form digital trunks of well-defined sizes. The data rate of any given span or the number of usable DS-0 channels it provides is standardized, but standards are different around the world. The most common forms of digital trunking facilities in use are the T-1 span (used in North America and Japan) and the E-1 span (used in Europe and elsewhere).

T-1

T-1 involves a facility, referred to as a *T-1 span*, that has a total capacity of 1.544 Mbps and supports 24 independent multiplexed channels. T-1 is typically implemented using four wires (two pairs). One pair is used to transmit data in each direction. T-1 also may be delivered through fiber-optic cable. The term *fractional T-1* refers to a normal T-1 span on which fewer than the full 24 channels are going to be used.

Figure 5-15. T-1 span

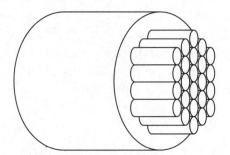

Each DS-0 channel on a T-1 span is equivalent to an analog line. In fact, T-1 was originally designed purely to concentrate the voice traffic from multiple analog lines. The 3.1 kHz audio bandwidth of an analog line is digitized and encoded into a 64 kbps media stream, which then is allocated to one of the T-1's channels. Unfortunately, because the voice media stream completely fills the DS0 channel and because each of the T-1 channels operates independently, signaling for calls is in-band. The in-band signaling approach used in T-1 is referred to as the *robbed-bits* method. This involves *stealing bits* from voice data and using them for encoding signaling information. Specifically, the standard T-1 method involves stealing every 48th bit (in each channel) for signaling. The signaling information is present at all times, but the lost bit results in an imperceptible loss of audio quality. The signaling bits are used to implement very simple signaling protocols that are based on the operation of the analog lines that the multiplexed channels supersede. For example, when a particular channel is not being used, the signaling bits are zeros in both directions. When the CO switch wants to present a call on a given channel, it changes the bits to be all ones. The CPE switch interprets the transition to all ones as a ringing indication, and it changes the bits it is sending to ones also in order to indicate that it wants to answer the call.

A limitation of the robbed-bits signaling method is that it limits the transmission of digital data using one of the DS-0 channels to 56 kbps in order to keep the digital data and the in-band signaling apart.

E-1

The *E-1 span* has a total capacity of 2.048 Mbps, which represents 32 DS-0 channels. E-1 uses out-of-band signaling, however, so 30 bearer channels are available for media streams and 2 channels are used for signaling and control. Because all signaling is out-of-band, each DS-0 supports digital data at 64 kbps. Like T-1, E-1 is typically available through either wire pairs or fiber optics.

Figure 5-16. E-1 span

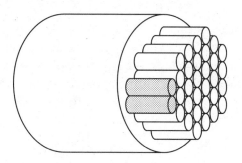

ISDN Primary Rate Interface (PRI)

ISDN Primary Rate Interface (PRI) provides ISDN signaling and services using T-1 and E-1 spans. Where T-1 is applicable, PRI consists of 23 ISDN B channels and one 64 kbps ISDN D channel for signaling. Where E-1 is applicable, PRI consists of 30 ISDN B channels and one 64 kbps ISDN D channel.

PRI offers two very significant advantages over basic T-1. The first is the use of a D channel for signaling. This means that each of the other channels in the T-1 span is free of in-band signaling and each is capable of carrying the full 64 kbps B channel media stream.

The second key advantage of PRI, which applies to both T-1 and E-1 spans, is that ISDN signaling provides a much richer feature set than the primitive, bit-oriented protocols that are used by basic T-1 and E-1. This additional functionality means, among other things, that individual B channels can be managed dynamically—in contrast to raw T-1/E-1 channels, which must have their use (incoming, outgoing, voice, digital data, etc.) predefined.

5.3.6. B-ISDN and ATM

Asynchronous Transfer Mode (ATM) refers to the next generation of digital networking fabric, which ultimately will eclipse the use of spans with channels of fixed bandwidth. ATM involves a form of time division multiplexing (see sidebar "Multiplexing" on page 261) in which the assignment of channels to time slots is managed dynamically. When a channel is established using ATM, the data rate and other QoS factors are used to determine channel assignment on the fly. Real-time isochronous traffic is guaranteed a consistent allocation of time slots, and asynchronous data is put into slots on an as-available basis. Each time slot in ATM actually is a packet of 53 bytes that carries with it the information needed to appropriately route it to its ultimate destination.

Broadband ISDN, or *B-ISDN* for short, is to ATM what BRI and PRI are to the digital network fabric based on fixed-size channels, that is, the set of specifications for performing signaling and for interfacing to the ATM portion of the digital telephone network.

The ability for ATM to supersede both existing telephony *and* telecommunications networks—by efficiently mixing all forms of telephony and telecommunications traffic without compromising the isochronous nature of telephony media streams—means that it represents the unifying technology in the world of wide area communications and networking.

5.3.7. DMT and ADSL

ADSL stands for *asymmetrical digital subscriber loop*. It is the best-known of a family of technologies, known as *DMT* or *discrete multi-tone*, that are designed to deliver multiple high-capacity digital channels using an existing single pair of analog subscriber loop wires. In DMT some of these channels are unidirectional, which is why the term *asymmetric* is used. These channels are referred to as "A" channels and support data rates of 1.5 Mbps. Asymmetric channels are seen as a vehicle for delivering TV programming or other media-intensive data streams to homes as an alternative to cable TV.

DMT may or may not emerge as the future of subscriber loop technology.

5.3.8. Cable TV Networks

Another subscriber loop technology that may emerge as the next dominant approach involves using a portion of the bandwidth available in cable TV networks.

Cable TV networks now connect a significant portion of homes and businesses throughout the world. This network has a great deal of capacity as it snakes through a neighborhood. While most of this capacity is used to carry TV signals in a single direction, a portion of this capacity can be allocated for carrying two-way media streams. By placing the appropriate transmitters and receivers at each home or business cable TV junction, this network can be used for both voice and digital data connections to the PSTN. This is illustrated in Figure 5-17.

Figure 5-17. Cable TV network providing subscriber loop

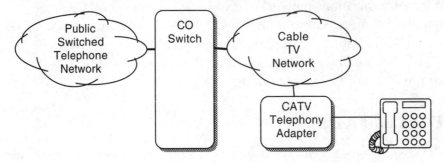

5.3.9. Wireless

Wireless telephony links come in many different forms, depending on the country involved and the nature of the application. They can be generally categorized as:

- Fixed or Roaming

    - Fixed wireless links involve some type of a fixed relationship between two transceivers.

    - Roaming wireless links involve a mobile transceiver that is able to connect with an appropriate nearby transceiver as needed.

- Dedicated or Shared

    - Dedicated wireless links are those that dedicate a wireless channel between a given pair of transceivers even when it is not in use.

    - Shared wireless links involve the sharing of wireless channels meaning that one transceiver may not be able to establish a link if another transceiver is using the channel.

Common wireless telephony facilities include:

- Cellular (AMPS and GSM)
- CT2
- 1.6–1.8 / 49.8–49.9 MHz cordless telephones
- 900 MHz cordless
- Radio telephones
- Aircraft public telephones
- Rail public telephones
- Microwave links
- Satellite links

The concept of a wireless local loop is illustrated in Figure 5-18.

Figure 5-18. Wireless local loop

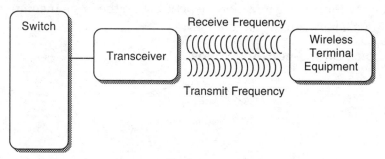

5.3.10. Internet Telephony

Another emerging facility for the transmission of telephony media streams is the Internet, as well as other equivalent local area network technologies. Delivering voice media streams in this fashion is referred to as *Voice-on-the-Net*, or *VON*.

All the lines and trunks we have examined so far have involved circuit-switched channels. This means that when a call has been established between two points, each leg of the connection is handled by a transmission facility that dedicates a particular portion of its bandwidth, or channel, to conveying the data stream. The result is referred to as a *switched circuit*.

With VON technology, in contrast, media stream channels are transmitted over networks where the bandwidth is shared between all the users of a given segment of the network. The network delivers information from one place to another using packets of finite size, which are addressed to a recipient. Anyone on the network can send and receive individual packets and in this way share the transmission facility. A telephony media stream in this environment is delivered as a sequence of packets from one point on the network to another. This is referred to as a *virtual circuit* because there is no dedicated bandwidth. The collection of VON endpoints that can be connected in this fashion is a VON network and is represented graphically as a cloud (as shown in Figure 5-19), much the same way that the other telephony networks are represented. Also as shown in the diagram, the scope of a given VON may be limited to just the Internet or just an intranet within an organization, depending on the implementation of the firewall between the two.

Figure 5-19. Voice-on-the-network clouds

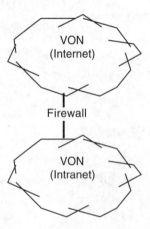

The trade-offs associated with packetized voice are considerable, especially as compared to circuit-switched voice. The significant trade-offs include:

- Audio quality is highly variable and connections are not reliable. The OSI network layer protocols that form the basis of the Internet and LANs are designed around the assumption that if there is network congestion (too many packets for the available bandwidth), packets will be buffered or thrown away. For conventional data streams, this means that packets will merely be delayed or retransmitted. For a VON media stream channel, this means that there is usually a break in the sound that is heard by a receiver.

- One of the benefits of Internet-type network technology is that networks can be built with many redundant paths between subnetworks, all of which may be used simultaneously. If one path becomes too congested, more packets start flowing over a different path. Unlike the voice traveling over a circuit-switched connection, which is isochronous, the VON media stream packets containing voice information can take different amounts of time to arrive at the destination.

- Shared bandwidth also means that there is less bandwidth available, so voice data generally must be compressed, resulting in some amount of sound quality loss.

- Unlike a switched telephone network, addresses on these types of networks are not necessarily fixed; there is no guarantee that a computer attached to the network at one point will be active or communicating at another point. To use VON, two parties must in effect agree to rendezvous unless one is guaranteed to be active at all times. VON therefore lacks the attribute of spontaneity associated with the facilities of the traditional telephone network.

285

On the other hand, VON offers some very compelling advantages to certain people:

- Calls placed on the Internet (or similar networks) are not billed as they would be in a traditional telephone network. Assuming someone is willing to put up with the quality trade-offs, conversations with points in virtually any location can be made effectively for free, assuming both have access to the Internet. (The cost of carrying the VON conversation is covered by everyone who shares in the cost of maintaining the network in question.) As bandwidth requirements grow, because this type of network use consumes considerably more bandwidth than conventional uses, the whole community sharing the network shares the costs of adding bandwidth.

- Multiple calls to the same destination can be more cost-effectively queued than if they had been dialed across a long-distance network (using an 800 number, for example) and then be queued at the destination while consuming expensive long-distance circuits. With Internet telephony, the call can be queued without consuming network bandwidth.

- Multi-point calls can be established by using broadcast and multicast mechanisms; this involves addressing the packets so they are delivered to a group of destinations. The result can mean inexpensive conference calls involving many people.

- Integration with other forms of data communication and published information may make the telephone call a secondary consideration.

- Call associated information (such as callerID, user data, and correlator data) is easy to implement and use if desired, without resorting to any form of in-band signaling.

The trade-offs with Internet telephony are considerable, but with the explosive growth of pervasive Internet access, the emergence of standards for VON media stream delivery, and the development of *Internet voice gateways*, many very appropriate uses of this technology are being identified. The best uses of Internet telephony include:

- Alternate trunks/lines for call centers

- Calling from an Internet telephone to a switched-circuit telephone

- Scheduled rendezvous of Internet phone users (i.e., conferencing)

5.4. Telephone Switches

CPE switching equipment can take many different forms. Historically, switching products have been divided into four principal categories based on differing underlying technologies and the resulting functionality. These four types are:

- Front-end switch

- Key system unit (KSU)

- Private branch exchange (PBX)

- Application-specific switches (e.g., ACD switch, etc.)

As all switch implementations evolved and came to be based on digital technology, features previously associated with only one type of device became easy to implement in any product, and the differences began to blur. Most new switch implementations are capable of virtually any feature, so the categorization of switch products has become increasingly arbitrary. Over time, the distinctions between switch types will continue to blur. Any switch may include any or all of the telephony resources described in Chapter 3.

The size of a switch is fundamentally measured in terms of two independent characteristics: its capacity for connecting lines (trunks and extensions) and its capacity for handling calls. A *non-blocking switch* is one that has enough capacity for calls that every line can be used simultaneously. A switch can vary in line capacity from just two lines to thousands of lines.

The other important characteristic that differentiates switch designs is their ability to be upgraded, and the limits to that upgradability. At the simplest level, a switch is simply a box, or *cabinet*, with a power supply and printed circuit boards. Switches that cannot be upgraded in any way have a fixed number of ports to which trunk and extension lines can be attached. A switch that can be upgraded may allow for the addition of entire auxiliary PBX cabinets, or for the addition of individual printed circuit boards to one or more *shelves* within the PBX cabinet. Add-in circuit boards may provide additional telephony resources of any type. Typically the cards will provide a set of additional trunk ports or extension ports of a given line interface type. Other examples of possible add-in functionality include media access devices (ranging from new DTMF detectors to a voice mail system), an ACD device function, an upgrade to switching resources (i.e., support for more conference calls), or a CTI interface.

5.4.1. Front-End Switches

A *front-end switch* is a very simple type of switch that sits between a series of CO lines (its trunks or network interface devices) and a corresponding number of telephone stations (its extensions or station devices). No switching takes place between extensions. The switch is capable only of establishing calls to connect (and disconnect) each extension to (and from) its designated trunk. A front-end switch may be modeled as illustrated in Figure 5-20. In the illustration, network interface devices are labeled with an "N," logical device elements are labeled with an "L," and physical device elements are labeled with a "P." Potential calls are shown in gray.

The main purpose for a front-end switch is providing a more sophisticated user interface to otherwise very primitive subscriber lines from another switch. This is accomplished by making the extension lines proprietary digital (or proprietary second pair) so that

Figure 5-20. Front-end switch model

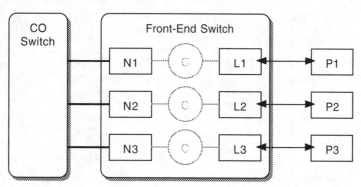

more sophisticated proprietary telephone stations can be used, or by providing a CTI interface on the front-end switch. The call processing function in the front-end switch accepts commands from the proprietary stations or the CTI interface and translates them into appropriate commands for the CO switch.

This type of switch is a very attractive way of providing Centrex services to a particular office or location. An appropriate front-end switch can be connected to a T-1 line from the CO switch, so that the telephone company need only provide four wires to an entire location of 24 or fewer users. Instead of accessing Centrex functionality through DTMF sequences that have to be memorized, Centrex telephone users at the location take advantage of sophisticated telephone stations, computer interfaces, or a combination of the two. (CTI configurations like this one are presented in Chapter 7.)

5.4.2. Key Systems

Key system units, or *KSUs* for short, are switches that are very similar to front-end switches, except that each logical station device has bridged appearances instead of the standard appearances seen in a front-end switch. Key systems use special telephone stations that are able to indicate the logical device with which they wish to interact. The typical mechanism used is a button, or "key," on the telephone set that a user can press. This is how the name "key system" was derived. A key system can be modeled as illustrated in Figure 5-21.

Figure 5-21. Key system model

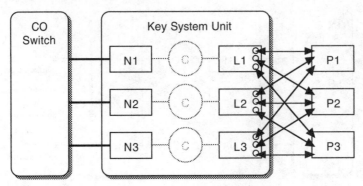

The bridged appearances used in a key system are shared–bridged by default, but a system that supports privacy implements exclusive bridging.

Some key systems have a special button (corresponding to a logical device) on the phone that is associated with an intercom or with the receptionist's telephone station. One example of this is shown in Figure 5-22. In this illustration, the receptionist's physical device is PR and it has access to logical devices L4, LR, L1, and L2. If the receptionist presses the button corresponding to L4, he is connected to the public address system or intercom and can announce something like, "Joe, the call on line 2 is for you." If anyone using P1 or P2 presses the button corresponding to L0, rather than connecting to a network interface device, the system places a call to LR, which then may be answered by the receptionist.

Like the front-end switch, the key system implements few, if any, telephony features and services itself. Instead, it treats its network interface devices as proxies and instructs the switch providing the corresponding line to carry out any requested services. In this way the key system adds the functionality of bridging (and, depending on the implementation, feature-rich telephone stations, and CTI interfaces) to whatever telephony features and services are supported by the switch that it is front-ending.

Figure 5-22. Example key system with attendant button and PA system access

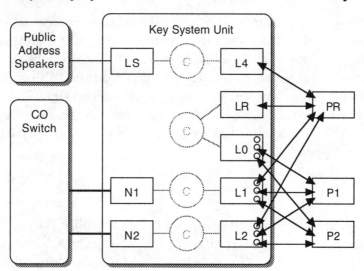

Squared System

A *squared key system* is one in which every telephone station has access to every logical device; typically the button for each of the logical devices is in the same place on every physical device. The key system shown in Figure 5-21 is squared. The term *squared* reflects the fact that, in this type of system, the maximum number of possible interactions that has to be supported via bridging is the square of the maximum number of physical devices supported. (It is assumed that there will never be more logical devices than physical devices.)

Hybrid

A *hybrid key system*, or *hybrid switch*, is one that supports devices which are both bridged and not bridged. If the devices with standard appearances can establish calls to or through devices other than a single designated network interface device, the switch is effectively a PBX that supports bridging.

Virtual Key System

In a *virtual key system*, there is no central KSU cabinet. Instead, pieces of the KSU functionality are implemented among each of the special telephone stations that make up the system. Each telephone station has a connection to every other telephone station and to all of the trunks. Call processing and switching functionality is distributed among all of the telephone stations. Despite the fact that there is no physical KSU cabinet, this type of key system is still modeled as shown in Figure 5-21.

5.4.3. PBXs

Private branch exchanges (*PBXs* for short) are also called *private automatic branch exchanges* (PABXs) or *computerized branch exchanges* (CBXs) by some vendors in order to contrast them with switching systems that require humans to perform the actual switching role (as was true with early key systems and cord boards). As manual systems are now all but obsolete, this clarification is not required; PBX is the preferred term for all products in this category.

A PBX is a general-purpose switch. It typically implements all telephony features and services internally, may connect any device to any other, and may support both standard and bridged appearances. A PBX may be modeled as illustrated in Figure 5-23.

Figure 5-23. PBX model

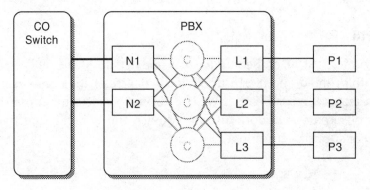

Distributed PBX

A *distributed PBX* is one that is implemented as a series of distinct PBX cabinets that may be located at some distance from one another. The implementation of call processing functionality and switching functionality is distributed among the pieces of the PBX; these pieces are constantly in communication in order to coordinate their activities. Some vendors support a feature in which, if a number of their PBXs are connected to one another in a private network using tie lines, the PBXs can establish digital data calls among themselves and use these to begin operating as a single distributed PBX.

Personal PBX

A *personal PBX* is to a traditional PBX what a personal computer is to a mainframe computer. The personal PBX is small, if not tiny by PBX standards. It has much less capacity than the average PBX, but is designed to be a peripheral in an individual's home, office, or small business. Unlike the traditional PBX, it is designed with an emphasis on requiring little, if any, administration or maintenance while delivering the complete range of telephony functionality found on a normal PBX.

Personal PBXs are designed to be used in environments that typically are equipped with a personal computer, so a full-function CTI interface is an important part of these products.

5.4.4. Application-Specific Switches

Application-specific switches are those that are not intended for general-purpose use in the same fashion as PBXs and KSUs, but instead are designed for some very specific and limited application.

The most common example of application-specific switches are stand-alone ACDs. These are basically PBXs that are built around a highly functional ACD or ACD group device. The network interface devices (trunks) for these switches are often connected to an organization's PBX, rather than to a central office switch. While many general-purpose PBXs include ACD or ACD group functionality, or have it as an option, other vendors specialize in ACD functionality. An application-specific switch is one way they can package their ACD functionality. (Another way is through a CTI interface.) One advantage—or disadvantage, depending on your point of view—is that application-specific switches may utilize proprietary telephone stations with displays that provide application-specific information.

5.4.5. Internet Voice Gateway

An *Internet voice gateway* is a switch with line interfaces in the form of one or more of the traditional switched telephone line interfaces described earlier in this chapter, and also in the form of VON media stream channels. The switched telephone lines typically are used as trunks; the VON media stream channels may be extensions or trunks, depending on how the Internet voice gateway is used.

As a switch, the gateway can place calls between the networks on either side, and can establish conference calls that connect multiple devices on either side. An Internet voice gateway typically is used to connect an intranet to a CPE switch, or to directly connect the PSTN to the public Internet. (Other combinations are equally likely, however.) This is illustrated in Figure 5-24.

Inbound Calls

One of the best uses of Internet telephony is allowing Internet users who are browsing World Wide Web information to talk to a human if they have questions, can't find what they're looking for, or want to place an order for something. This is illustrated in Figure 5-25.

This is an example of good Internet telephony use, because many residential Internet use all their available connectivity to make their connection to the Internet and would have to disconnect in order to place the phone call. In addition, because the Internet telephony call is being placed to a well-known location that is expecting such calls (i.e., the organization managing the Web site), the issues of spontaneity don't apply.

Figure 5-24. Internet voice gateways

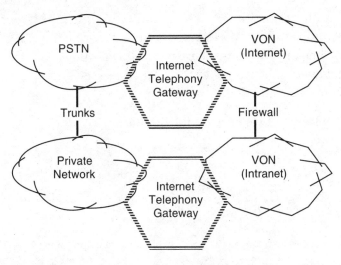

Figure 5-25. Using Internet telephony for inbound calls

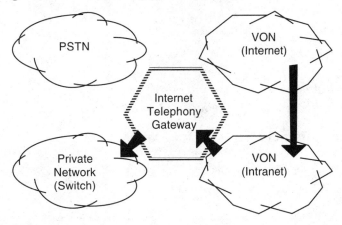

In this scenario, the Internet telephony gateway is effectively a trunk as far as the rest of the called organization's telephone system is concerned. The difference is that correlator data associated with the incoming call may be used to link the call to the Web site context. The Internet voice gateway allows existing call center agents to handle Internet telephony calls in addition to their normal calls, which are using existing telephone station equipment.

Internet Tie Lines, FX Lines, and OPX Lines

Another application of Internet telephony through an Internet voice gateway involves alternative implementations of tie lines, FX lines, and OPX lines in situations where the reduced quality and reliability of Internet telephony are not of concern. These uses are illustrated in Figures 5-26, 5-27, and 5-28.

Figure 5-26. Using Internet telephony for tie lines

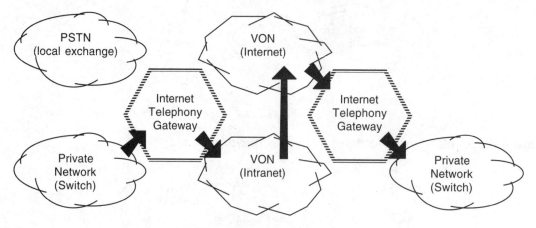

A tie line between two CPE switches is accomplished by routing calls between Internet voice gateways attached to each.

Figure 5-27. Using Internet telephony for FX lines

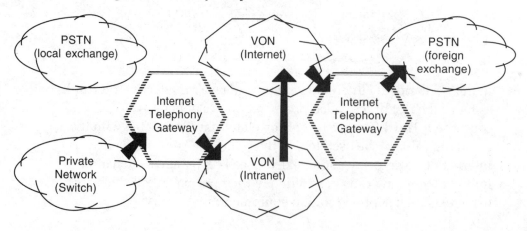

An FX line can be implemented by working with an alternate wire line provider in the foreign exchange area that can provide access to the local switch in that location through available trunks on its Internet voice gateway.

Figure 5-28. Using Internet telephony for OPX lines

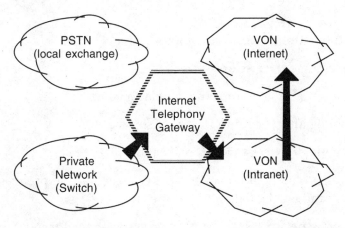

The Internet telephony approach to an OPX line is used when, for example, an employee who telecommutes is using all available connectivity for Internet access. This person will be at a fixed network address for an extended period of time, so Internet telephony calls may be directed to that person's machine on the Internet.

5.5. Switch Peripherals and Add-Ons

A number of dedicated peripherals are available for most switches. These peripherals generally are dedicated products that provide service through one of the switch's interface ports, by attaching as a station device, or a combination of the two.

5.5.1. OA&M Interfaces

A switch's *operations, administration, and maintenance,* or *OA&M interface*, provides the functionality necessary for setting up and administering it. The OA&M interface may also provide diagnostic or status information and allow telephony resources to be taken in and out of service.

Standard functions of an OA&M interface include:

- Assigning device addresses;

- Assigning bridged appearance relationships;

- Defining and assigning classes of service;

- Assigning devices to hunt, pick, and ACD groups;

- Setting up dial plan management rules, including prefixes, LCR rules, and associated network interface devices;

- Defining network interface device associations for incoming calls on non-DID trunks;

- Fixing system defaults for forwarding; and

- Configuring media service devices.

Once a switch has been set up, the primary use of the OA&M interface is for *moves, adds, and changes,* or *MACs* for short. This refers to the day-to-day need to cope with people moving from office to office, and joining or leaving an organization. The assignment of telephone numbers to lines must be modified every time one of these events takes place.

An OA&M interface implementation not only permits administration of these features, but also exposes a mechanism by which a human user can interact with it. Three common approaches are providing a console interface, providing support for the direct attachment of an ASCII terminal, and using a LAN-based connection to a separate computer.

Console

A commonly implemented OA&M mechanism for small switches involves using telephone stations.

If the OA&M options are sufficiently simple, an ordinary touchtone phone can be used. In this case, a special sequence of DTMF tones is interpreted as a command to place the telephone station into OA&M console mode; then more DTMF tones can be used to set features. While this type of implementation is inexpensive, it also is error-prone and difficult to use.

Another variation on this approach involves special console telephone stations with many buttons and a large display. When this telephone station is placed in OA&M mode, instructions and feedback appear on the set's display and buttons are used to select options and make settings.

Terminal Interface

The most common access mechanism for OA&M implemented in PBXs is the OA&M interface serial port. This is a serial port (or set of serial ports) that allows a terminal to be connected, either directly or remotely through a modem. Once the terminal is connected, the OA&M interface can be manipulated as if it were any other type of text-oriented computer to which a terminal had been attached. Commands are entered on the terminal keyboard and feedback appears on the screen. Some vendors implement a very primitive user interface based on a simple stream of characters that are read in and printed out. Other vendors provide a slightly easier-to-use interface that employs the whole surface of the terminal display, making status easier to read and understand.

Network Interface

The third approach, which is likely to become the dominant one eventually, involves using a LAN connection so that application software on a computer can be used to connect to the switch for purposes of monitoring and configuring it.[5-10]

5.5.2. Telemanagement Systems

A *telemanagement system*, sometimes referred to as a *call accounting system*, is a computer-based system designed to help manage and account for the use and assignment of a switch and associated facilities. Telemanagement systems connect to the switch's accounting interface to obtain information pertaining to the usage of telephony resources.

The switch's accounting interface is typically connected to the telemanagement system using an RS-232 serial port. The information delivered using this serial port is often referred to as *call detail recording* (*CDR*) or *station message detail recording* (*SMDR*) information. For each call originated, the accounting interface generates a record of information detailing the starting and ending date and time, plus all the call associated information including the calling device, the called device, the network interface device used, the account number and access code used, etc. This information streams out of the switch through the appropriate serial port. Some organizations don't use this information and leave the port unconnected; others attach a line printer so that this information is captured and later can be analyzed manually. In general, however, most organizations want the ability to make full use of this information, so they install a telemanagement system.

5-10. OA&M protocol — Though a specific OA&M protocol for switches does not exist at the time of writing, this is a likely target for industry groups now that standard CTI protocols have been developed. Many vendors have already begun to build custom extensions to SNMP, which is the TCP/IP-based management protocol that is the de facto standard for network management. This is a good place to start and is likely to be the basis for any future OA&M protocol.

Telemanagement software typically uses the CDR information received from the switch, along with all the rate information for the carrier associated with each network interface device. The telemanagement system estimates the cost of each call and can provide reports that can be used in various ways, including:

- Identifying toll fraud activity;

- Billing clients;

- Billing against projects or departments;

- Validating telephone bills from carriers; and

- Adding a telephone charge to a hotel guest's bill.

Telemanagement systems also are used to conduct traffic studies. A *traffic study* is a statistical analysis of call data to determine whether telephony resources and systems are being used in a cost-effective or optimal fashion. For example, a traffic study might determine that even in peak periods only 75 percent of the trunks on a switch are ever used. With this information, the organization can save money by decommissioning the unneeded trunks. Another example might be the discovery that most calls being delivered to a call center are not arriving with ANI information. In this case, the call center's software can be adjusted to deal with ANI-free calls more optimally, and the carrier can be contacted to see if the problem is in their network.

5.5.3. Voice Mail

A *voice mail system* is a switch peripheral that uses media access to record and play back messages. Voice mail systems can take many different forms, but they typically have the following basic features:

- Call associated information, such as the identity of the called device, is used to determine the intended voice mail recipient.

- If call associated information is unavailable, the voice mail system may interact with a caller to obtain the identity of the desired voice mail recipient.

- Based on the identity of the recipient, a custom greeting is played. This might be simply the recipient's name incorporated into a single, system-wide greeting. It also could be a prerecorded greeting left by the recipient, or one of a set of prerecorded recipient greetings that is selected based on a rule such as the time of day.

- Messages may be tagged with a priority (urgent/not urgent), a status (new/listened-to/saved), and time and date information. When the messages are retrieved, they are accessed based on this tagging.

- Tagging can be used to determine how a recipient should be notified of messages. For example, a system might flash a special "message waiting" lamp on a telephone set to indicate new messages, and might send a pager message to indicate the arrival of new/urgent messages.

- Message tagging may be used as a basis for deleting certain messages automatically. For example, all messages older than two weeks may be deleted automatically, or all messages that have been listened to (but not tagged as saved) may be discarded.

- Recipients can retrieve messages by calling a special number and using DTMF commands to authenticate themselves and to interact with the voice mail system. Typical commands include next message, previous message, delete message, save message, rewind, fast forward, and pause. When messages are played back, all or some of the tag information may be communicated as well.

Voice mail systems may be attached to a switch in many different ways. Some voice mail systems are designed to be added directly into a switch, and all logical devices used by the voice mail system are strictly media access devices. Other voice mail systems use lines from the switch, and their logical devices are actually station devices that have associated media services.

5.5.4. Universal Mailbox

A *universal mailbox system* is a voice mail system that typically has the following enhancements:

- Accepts faxes in addition to voice messages;

- Is integrated with an electronic mail system in some fashion; and

- Can be accessed electronically through a computer interface.

A universal mailbox system generally allows a recipient to view a list of all of the items in his or her mailbox, and allows individual messages to be retrieved through a computer interface.

Universal mailbox system implementations generally have a connection to a local area network for computer access, in addition to their connections to the switch.

5.5.5. UPS

One of the most important switch peripherals is an uninterruptible power supply (UPS) that will ensure that the telephone switch will continue to operate in the event of power failures or other power anomalies. Aside from the fact that a switch often is a significant asset, it is generally one of the most relied-upon systems for any organization. A UPS will protect the switch from power sags and spikes that have the potential to do serious damage. In the event of a power failure, the batteries (or generator) associated with the UPS will ensure that the telephone system continues to operate. In prolonged power outage, the switch can be shut down in a graceful fashion automatically.

5.6. Telephone Stations

Telephone stations are the telephony products that are connected to switches using line interfaces corresponding to the switch's station devices. Telephone station design is limited only by the imagination of the product designer. Thousands of different telephone designs have been developed over the last century. A number of representative examples are presented in this section to demonstrate how easily all of these products can be modeled; these examples are by no means exhaustive, however. Expect to find any combination or permutation of the products described here.

In Chapter 3, section 3.6 we saw how the physical element of a station device is modeled, and in Chapter 3, section 3.9 we saw how physical device elements may be associated with one or more logical device elements through device configurations. This section puts these concepts to work as it describes a range of typical telephone station designs.

5.6.1. Single-Line Telephone

The device configuration for a simple, single-line phone can be modeled as illustrated in Figure 5-29. This example represents a basic POTS telephone set, with only the most basic features, on a dedicated line.

Figure 5-29. Dedicated POTS line station example

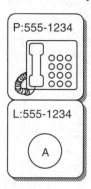

In this case the device configuration consists of a single device with a physical element and a logical element. (Both elements have the same label.) The logical device element has non-addressable standard appearances. The physical element consists of:

- A ringer

- A hookswitch that is controlled locally

- An auditory apparatus in the form of a standard handset with a single, fixed-gain microphone and fixed-volume speaker

- A dial pad with twelve buttons

The telephone station itself might look like the one pictured in Figure 5-30.

Figure 5-30. Simple POTS telephone set

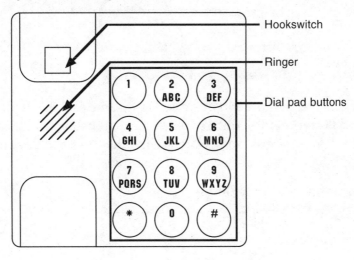

5.6.2. Bridged Line

When multiple analog telephone sets are bridged physically onto the same analog line, the result is sometimes referred to as a *party line*. The logical device corresponding to the line behaves as described for interdependent–shared–bridged appearance (see Chapter 3, section 3.8.3). This device configuration for a POTS telephone station is shown in Figure 5-31.

Figure 5-31. Bridged POTS line station example

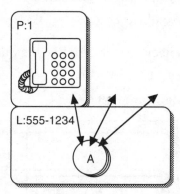

The physical element used in this case is the same as the one for the dedicated POTS line. The difference here is that the logical device element is simultaneously associated with other physical device elements as a part of their device configurations.

Figure 5-32. Key phone station example

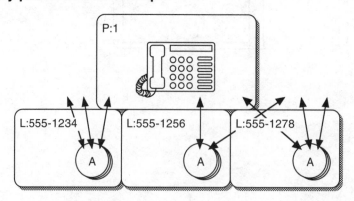

5.6.3. Key Telephones

A *key telephone* is a special phone designed to be used with a particular key system (as described in section 5.4.2). Simple key telephones, like the one used in this example, are essentially the same as the physical elements used in the last two examples, except that they have extra buttons (so-called *line buttons*) to allow access to multiple logical devices using bridged appearances. The device configuration for a key telephone station is illustrated in Figure 5-32 and the appearance of a corresponding telephone set is depicted in Figure 5-33.

Figure 5-33. Simple key telephone set

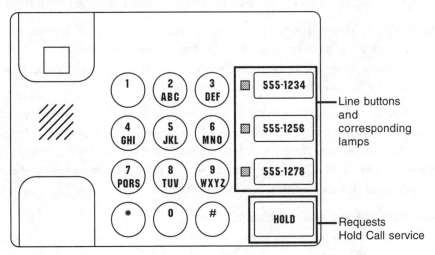

Key telephones typically use one or more lamps to indicate the presence and status of a call associated with a given button. For example, in one implementation the following indications might be used:

- Lamp mode: *off*
  There is no associated call (*null* connection state).

- Lamp mode: *steady*
  There is a connection in the *connected* or *initiated* state.

- Lamp mode: *wink*
    There is a connection in the *hold* state.

- Lamp mode: *flutter*
    There is a connection in the *alerting* state.

- Lamp mode: *broken flutter*
    There is a connection in the *queued* or *fail* state.

5.6.4. Multiple Line Telephones

Multiple line telephone stations that make use of dedicated—not bridged—logical devices come in two varieties: analog and digital.

An analog multiple line telephone station connects directly to multiple lines, so each of the associated media stream channels are simultaneously dedicated to the telephone station, but only one can be used by the telephone station at a time.

The digital version is often called a *multiple DN telephone* (for *multiple directory number*). In this case the telephone station is connected only to a single line, but the media stream channels supported by the line are associated with whichever logical device element is interacting with the physical device element at a given instant.

Figure 5-34. Multiple line station example

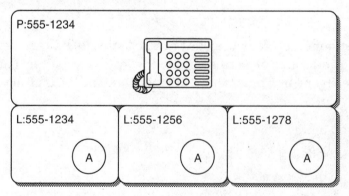

The telephone station itself generally appears the same as in the key telephone shown in Figure 5-33. The logical device elements in the device configuration, however, have non-addressable standard appearances. Figure 5-34 illustrates a multiple line device configuration in which the logical device labeled "555-1234" is part of the base device for the device configuration.

5.6.5. Multiple Appearance Telephones

Multiple appearance telephones are found on digital lines. They are like multiple line stations. Rather than having buttons representing different logical devices, however, they have buttons representing different addressable appearances associated with the same logical device. The functionality for the telephone user is therefore very similar, despite the fact that the implementation is quite different. The appearance behavior may be basic or selected. If it is basic, a new call will be presented in the *alerting* state at each of the appearance buttons, and any of the buttons may be pressed to answer it using the corresponding appearance. If the appearance behavior type is *selected*, new calls will appear only at one of the appearances. A *multiple appearance* device configuration is shown in Figure 5-35 and the corresponding telephone station is shown in Figure 5-36.

Figure 5-35. Multiple appearance station example

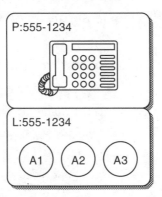

P:555-1234

L:555-1234

A1 A2 A3

Figure 5-36. Multiple appearance telephone set

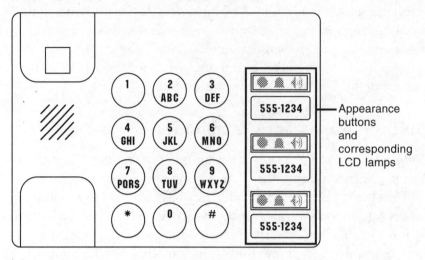

Note that the telephone set in this example uses lamps in yet another way. In this case there are three distinct lamps for each appearance button. These lamps are in the form of symbols (implemented using LCDs for each button) that can be turned on and off. These are used as follows:

- No symbol showing
     There is no associated call (*null* connection state).

- Speaker symbol
     There is a connection in the *connected* or *initiated* state.

- Bell symbol
     There is a connection in the *alerting* or *queued* state.

- Stop sign
     There is a connection in the *hold* state.

5.6.6. Assistant's Telephone

In many workplace environments, an assistant handles calls on behalf of his or her manager. The logical device associated with the manager's telephone number therefore must be bridged, while the assistant's need not be. This is one case of a *hybrid* device configuration. An example consisting of an assistant's telephone with two managers and multiple appearances for the assistant's logical device is shown in Figures 5-37 and 5-38.

Figure 5-37. Assistant's telephone set

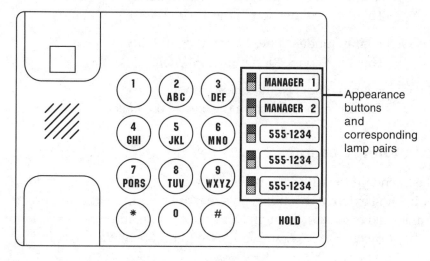

Figure 5-38. Assistant's hybrid station example

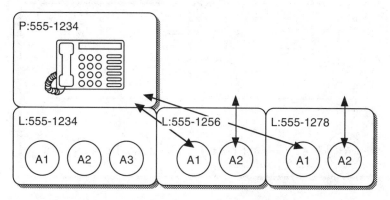

This example illustrates yet another way that lamps may be used in a given design. In this case, each appearance button is associated with two lamps, one red and one green. This implementation might use these lamps as follows:

- Green lamp mode: *off*, Red lamp mode: *off*
  There is no associated call (*null* connection state).

- Green lamp mode: *steady*, Red lamp mode: *off*
  There is a connection in the *connected* or *initiated* state.

- Green lamp mode: *flutter*, Red lamp mode: *off*
  There is a connection in the *alerting* state.

- Green lamp mode: *off*, Red lamp mode: *steady*
  There is a connection in the *hold* state.

- Green lamp mode: *off*, Red lamp mode: *flutter*
  There is a connection in the *queued* or *fail* state.

5.6.7. Attendant Console

The attendant for a PBX switch is responsible for answering and redirecting external incoming calls that are not automatically directed to a desired extension through DID, DISA, or an automated attendant. In the *attendant console* example shown in Figure 5-39,

Figure 5-39. Attendant console station example

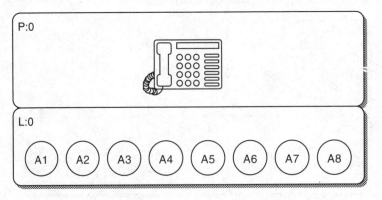

the station's logical device element has addressable selected–standard appearances that correspond to each of the six network interface devices (trunks) that the attendant is responsible for answering. Two additional appearances are used for internal calls. All external incoming calls associated with one of these trunks are presented to the appropriate corresponding appearance. The attendant can then answer the call, find out the desired destination, and redirect it.

One rendition of the corresponding attendant console telephone set is depicted in Figure 5-40. In this example, the attendant console has four types of buttons:

- Dial pad buttons

- Function buttons ("hold," "transfer," "drop," "park")

- Appearance (trunk) buttons

- Single step transfer (extension) buttons

Figure 5-40. Attendant console telephone set

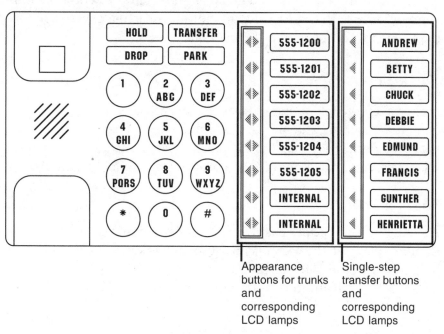

Appearance buttons for trunks and corresponding LCD lamps

Single-step transfer buttons and corresponding LCD lamps

The trunk buttons are the call appearance buttons associated with the trunks. Extension buttons are special-function buttons that perform a *single step transfer call* service to a preprogrammed extension number. In normal operation, the attendant simply presses a trunk button to answer an alerting call and then presses the extension button to transfer it to the requested destination. The dial pad and function buttons provide additional flexibility, such as selecting reach extensions that are not listed. Pressing the "transfer" button in this example performs a *consultation call* service, with *transfer* as the consult purpose. The desired extension number is then entered on the dial pad, followed by a press of the "drop" button. This completes the transfer operation by requesting the *transfer call* service followed by the *clear connection* service. Similarly, the "park" button invokes the *park call* service and allows the attendant to park a new call at an otherwise busy extension.

It is useful for the attendant to be able to see what extensions are busy before attempting to transfer a call. Therefore every trunk and extension button is provided with an "in-use" lamp, in addition to the lamp used for each trunk button to indicate an actual connection for the attendant console in the *alerting* state.

5.6.8. Desk Sets

Desk sets refer to telephone stations that are attached to a wire (tethered to a desk) and are used by someone on a day-to-day basis at his or her desk. They may have an analog telephone line interface, an ISDN telephone line interface, or a proprietary digital line interface.

Speaker Phone

A *speaker phone* typically has two auditory apparatuses. The first is the standard handset; the second consists of a speaker and a microphone built into the base of the telephone station. This is illustrated in Figure 5-41. Each auditory apparatus has its own hookswitch.[5-11]

Figure 5-41. Speaker phone

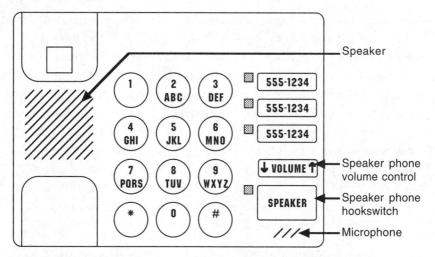

Speaker-only Phone

A *speaker-only phone* is a cost-reduced version of a speaker phone where the microphone in the base is omitted; the media stream associated with a call can be heard, but only silence is sent in the other direction.

Wireless Handset

Another variation on desk set design involves wireless handsets. The handset for a telephone station may be connected to the base using a wireless connection rather than the traditional coiled cable. The wireless handset could be provided in addition to or instead of a base handset.

5-11. Internal hookswitch — In many speaker phone implementations, the actual hookswitches are internal to the telephone station and are controlled in such a way that only one is active at a time. The cradle hookswitch control and the speaker hookswitch control are used as inputs to reset the actual hookswitches.

Headsets

Those who spend prolonged periods using a telephone generally take advantage of yet another type of auditory apparatus: the headset. Headset designs range from the bulky, dual-cup model with a big microphone boom that drops down in front of the speaker's mouth, to a tiny unit that combines a speaker and microphone into a single ear piece that rests in the ear. Those in the first category are good in noisy environments; those in the latter are good wherever they are practical. Headsets generally attach to desk sets in place of or in tandem with a handset. Desk sets that are designed with headsets in mind use a hookswitch button on the telephone set, rather than a cradle hook switch. If both a handset and headset are attached to the handset connector, a switch is provided to switch between the two.

Headset-only telephone stations are used in environments, such as call centers, where jobs are telephone-centric. If CTI is used in such an environment, the computer interface provides all of the call control; the telephone station itself is used only for the auditory apparatus (here a headset-only unit), or for dealing with a call already in progress in case the computer fails for some reason. Such a device is pictured in Figure 5-42.

Figure 5-42. Headset telephone set

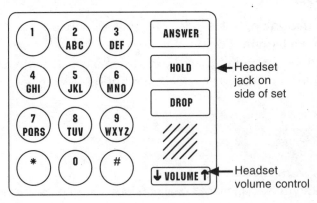

Wireless Headsets

A variation on headset design is using a wireless link[5-12] between the headset and the telephone station to allow more mobility for the headset user. In any office environment, this allows individuals to leave their offices for any reason and not have to "detach" from their telephones. It also means that they can be notified of incoming calls at their telephones while away from their desks. Wireless headsets are very useful for call center agents who must research items on behalf of callers and must roam away from the desk. For example, if a caller to a catalog clothing company requests a jacket the same color as the one he bought several years ago, the agent handling his call might have to walk to a bookshelf with all of the archived catalogs. With a wireless headset, the agent can walk to the shelf and look up the old jacket while continuing to talk to the caller. Without the wireless headset, the agent would have had to put the caller on hold, take off the headset, check the old catalog, and then reverse the process.

Display

A display is a matrix of characters that can be displayed by a telephone station. The actual matrix managed internally by the telephone can be larger than the visible portion, provided that buttons on the station allow for scrolling in some fashion. The display may show characters in any character set, using LCD or LED bitmaps or by forming the character in each position using a multi-segment approach.

Though the two most common uses of a display are to allow on-hook dialing and to show callerID information, there is no limit to the type of call associated and device associated information that can be presented. For example, some telephone sets use the display to implement a type of help function that steps a person through the use of a particular feature.

5-12. Wireless headset technology — Spread-spectrum radio frequency technology is preferred because it is least susceptible to interference from other wireless headset users in close proximity.

Figure 5-43. Display phone

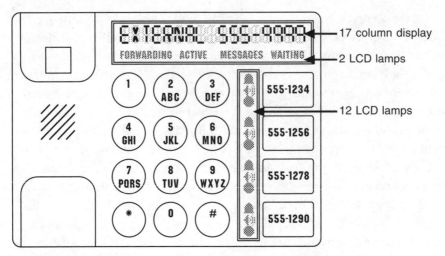

The display area should not be confused with other areas of a telephone set's surface that may use LCD as a technique for implementing lamps. In Figure 5-43 there is only one display area, which is one line high and 17 characters across, but it also uses a portion of the same LCD as a message-waiting indicator and a lamp that indicates that forwarding is active.

5.6.9. Wireless Telephones

Wireless telephone stations are those that connect to a switch through a wireless "line" rather than a cable of some sort (see section 5.3.9). Wireless stations may connect to a public switch, or to a PBX, or in some cases to use a public switch automatically if its home PBX is not accessible.

Wireless telephone stations tend to have many of the same features that apply to desk sets, but are constrained by the fact that they must be as small as possible.

Display

A display on a wireless telephone station is another essential element. The need for a display is driven primarily by the need to support on-hook dialing. The telephone set's user can enter a number, backspace to correct mistakes, etc., and only press the send button (which requests a *make call* service) when the number is correct. As with desk sets, displays are also used for presenting callerID information (where available) and other implementation specific help information.

Speed-Dial Buttons

Wireless phones tend to have many (100 or more) speed-dial or "rep" buttons (i.e., stored numbers). Because the units are small, however, there isn't room to have each of these buttons on the surface of the phone. Instead, these speed-dial buttons are virtual, that is, they exist only inside the phone. Users of the wireless telephone set press a sequence of physical buttons in order to trigger a "press" of the virtual speed-dial button of interest. A CTI interface to the same phone would allow using the *set button information*, *get button information*, and *press button* services, respectively, to program a given button with a speed-dial number, retrieve the number, and dial it.

Lamps

Examples of lamps used on cellular phones include those for indicating battery levels, the presence of an active connection, travel outside the local service area, and the lack of any available switch providing service.

Ringer

The ringer in a cellular phone may be auditory or vibratory, or there may be one for each. (Indicating ringing through vibration is decidedly not a feature found on desk sets.)

319

Auditory Apparatuses

Wireless telephone stations are typically in the form of a handset, so the built-in auditory apparatus is a handset. There usually is a volume control for the associated speaker.

In addition, most new wireless telephones support an earphone jack designed for the direct connection of an all-in-one headset. This is very useful in nearly every context where a cellular phone may be used, but it is particularly important in vehicles because it allows the cellular phone to be used while keeping both hands free for driving.

A speaker phone auditory apparatus is an alternative to the headset for use in vehicles. It involves installing a speaker and a microphone in appropriate locations in the car.

5.6.10. Multi-Function Telephone Stations

Many telephone station products also have other built-in functionality. These *multi-function telephones* have both a telephone station portion and a separate portion that might have related or unrelated capabilities. Examples include:

- Clock-radio telephones

- Pay phones with attached data terminals

- Fax phones

- Video phones

- Combined phone and answering machine devices

In the cases where the other portion of the product's functionality relates to media access, the media service instance is treated as part of the given device and is associated with the its logical element. The media service resources in these devices share the same appearance as the physical device. For example, a fax phone consists of both a physical element (the telephone station portion) and a logical element that can, when requested, transparently associate the fax media service instance with calls. When it does so, the auditory apparatus on the associated physical element may or may not be muted during any subsequent fax transmission. In this example, the device configuration is the same as for the POTS telephone stations described earlier (see Figure 5-29).

5.7. Telephone Station Peripherals

Telephone station peripherals are devices that can be connected to a telephone line in addition to a telephone station (through physical bridging or tandem connection), or instead of a telephone station. As with telephone stations themselves, the possibilities in this category are endless; the examples presented here are merely representative.

5.7.1. CallerID Displays

CallerID displays are small units with LCD displays that can be bridged onto a telephone line (typically an analog telephone line). They display the callerID information associated with each incoming call (assuming that callerID service is subscribed to). This type of product is an alternative to replacing an entire telephone set with one that has a display.

5.7.2. Call Blockers and Call Announcers

Call blockers and *call announcers* are variations on callerID displays. While they might or might not have a display for showing callerID, they are connected to the phone line coming into an office or home in tandem with the telephone(s), and they use the callerID information associated with a call to selectively block and/or customize the announcement of calls. A call blocker can be programmed with certain numbers that are to be blocked, that is, if calls are presented with one of these numbers, no telephones attached to the blocker will be rung. The blocker might or might not play a message to the caller. Similarly, a call announcer may use the callerID information to ring the phone in some customized fashion. These devices are functionally equivalent to a telephone company's call blocking and custom ringing services.

5.7.3. Media Access Products

Media access products are products that attach to computer equipment in some fashion in order to get access to one or more media stream channels on a telephone line.

External Modems

External modems are products that can be attached externally to a computer and typically are connected through a serial cable or serial bus. External modems allow for forms of media access including modulated data, modulated fax data, and sampled audio data. An external modem may be attached to one or more telephone lines.

Data Units and Terminal Adapters

Data units and *terminal adapters* are the digital data equivalent to modems on a telephone line. They allow computer systems to get access to the digital data media streams carried by a particular telephone line.

Media Interface Cards

Media interface cards are add-in cards designed to be placed inside a computer system. They attach to one or more telephone lines and implement any type of media access, ranging from speech recognition to modem functionality.

Software Modems

Software modems implement fax and data modulation as a modem would, but do so in software running on a computer system. The software modem works with media streams that are delivered to and received from a telephone line using one of the pieces of hardware described above, or through a simple piece of hardware that looks like a modem but acts only as a digitizer and transducer of the analog signal on an analog line.

5.8. Review

In this chapter we have explored the services, products, and technologies that can be combined to build a telephone system. In particular, we have seen how the tangible services and products offered by telephone companies and equipment vendors are all easily abstracted using the concepts covered in Chapters 3 and 4. This is what makes CTI possible, as we will see in the remainder of this book.

The telephony features and services available to a given individual are a function of the features, services, and channels provided by one or more telephone companies, combined with the features and services provided by the individual's *CPE* or *customer premises equipment*. The CPE might range from a single telephone station to a network of CPE switches, each with many associated telephone stations.

A *telephone company* provides connectivity from a *central office switch* (or *CO switch*). In addition to providing media stream channels, the CO switch provides certain telephony features and services that are *subscribed* to from the telephone company. Media stream channels and signaling are carried over facilities, generally called *lines*, that include (but are not limited to) *analog, ISDN Basic Rate Interface* or *Primary Rate Interface (BRI or PRI)*, *T-1/E-1 span, B-ISDN/ATM, CATV, Internet telephony*, and various forms of *wireless* links. Digital lines use *multiplexing* technology to combine multiple signaling and/or media stream channels onto a single set of cables or a single wireless link.

A *telephone switch* is a set of telephony resources connected to one or more other switches with trunk lines that correspond to network interface devices and extensions, or subscriber lines, that correspond to its station devices. CPE switches include *front-end switches, key system units (KSUs), private branch exchanges (PBXs), application-specific switches*, and *Internet voice gateways*. PBXs generally implement a complete set of telephony features and services, while traditional KSUs and front-end switches simply connect and disconnect telephone stations from trunks and take advantage of telephone features and services supplied by the CO switch to which it is connected.

Telephone stations and *telephone station peripherals* are implementations of telephony resources that connect to lines from switches. Telephone stations generally rely on their switch for telephony feature and service implementation. The most common line interface types used by telephone station equipment are analog, ISDN BRI, proprietary digital, and wireless. Telephone stations include a physical device element that provides at least one auditory apparatus for accessing the media stream of a call. The use of lamps, buttons, ringers, and a display in the design of a telephone set varies dramatically, as a result both of the device configuration(s) and of the aesthetic, cost, and usability trade-offs made by product designers.

6.
CTI Concepts

The one telephony resource that has not yet been discussed in detail is the CTI interface. This telephony resource is the one that allows a particular telephony resource set, or telephone system, to offer CTI functionality and become part of a CTI system. In this chapter we will explore what CTI interfaces are, how they work, and what functionalities they provide.

As we have seen already, most telephone systems offer, or are capable of offering, a very rich suite of functionality. The vast majority of people traditionally use only a fraction of the functionality that their telephone system offers, however, because that functionality is locked away behind a difficult-to-use telephone keypad interface. The CTI interface represents an alternative means of reaching this functionality with all the power of a modern computer-based user interface design.

6.1. CTI Abstraction

An abstraction is a myth that we create in order to make something manageable. In the case of CTI, the abstraction of telephony functionality makes it possible to design software that will work with more than one telephone system and support more than one user's paradigm.

The huge diversity in telephone system implementations means that there is an equally huge diversity of very different implementations. In the past (before the development of a robust abstraction of telephony functionality), software developers, installers, and customers who wanted to build CTI systems had no choice but to be aware of the internal designs and proprietary terminology, concepts, rules, and behaviors of every telephone system.

The development of a universal abstraction allows any implementation, regardless of its size, to be described in the same terms. One way to think of this is that the telephone system presents a *façade* that appears to all observers to behave precisely as the universal abstraction dictates. Behind the façade, however, the telephone system is doing whatever is appropriate to translate between its own internal representations and those in the façade. To the observer of the façade, there is no difference between this telephone system and one with the same functionality that might be built with an internal representation based directly on the abstraction. This is illustrated in Figure 6-1. This translation between the universal abstraction (the façade) and the actual implementation is the role of the CTI interface.

6.1.1. Observation and Control

As we have already learned, computer telephony integration allows computer systems to both observe and control resources and entities in telephone systems.

This observation and control is not directly a function of the actual telephone system implementation, but of the abstraction or façade that the telephone system presents through a CTI interface. The internal

Figure 6-1. Telephony abstraction is a façade

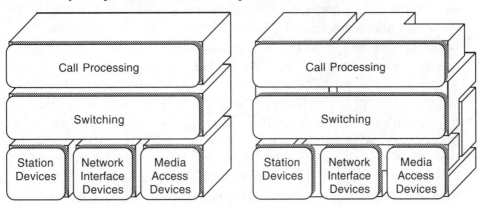

implementations of a PBX handling an incoming call from a central office switch on a T-1 trunk and presenting it to a DID extension is very different from the internal implementation of a simple POTS telephone receiving a call on an analog line—but two computer systems using CTI interfaces to monitor each of these station devices would see the same thing through the abstraction: a new call being presented to the device in the *alerting* state.

Generally speaking, observation and control are closely linked. While there are some exceptions, virtually every useful application of CTI technology needs the ability to observe activity within the telephone system so that it can control it in some desired fashion or, at a minimum, provide logging and monitoring functions. In other words, the computer system has to be able to see what it is doing before it can manipulate resources and objects within the telephone system. Furthermore, because the telephone system is a dynamic, constantly changing facility in which every component has a state or status that affects what can and cannot be done, the observation of the telephone system must be continuous and ongoing.

6.1.2. Manual versus CTI interfaces

A good way to think of a CTI interface is as an alternative to the standard observation and control interface used with telephone systems, i.e., the telephone set.

In the example illustrated in Figure 6-2 a CTI interface is being used by a computer to observe and control all of the activity associated with a particular telephone. Like the human sitting next to the telephone, the computer can "see" all of the lamps that are lit, the buttons that may be pressed, the text on the display, etc. Just like the person using the telephone, the computer can place calls, answer calls, press buttons, etc. Bounded only by the capabilities of the CTI interface in question, the computer can, in fact, do anything the human can do with the telephone—and possibly more. It is as if the computer can reach out an invisible electronic arm and do the same things that the human can do to the telephone set. In this example, both the computer and the human have free access to the telephone set and can manipulate it independently.

Figure 6-2. Multiple interfaces to telephone functionality

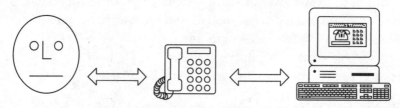

This example illustrates an important point about the way CTI interfaces generally work. To the extent that the CTI interface allows observation of a particular device, a computer may be just one of many observers (either humans or other computers) of that device. To the extent that the CTI interface allows controlling the device, that control is not exclusive. The computer cannot prevent a human from pressing a particular button, lifting the handset at a given instant, or performing some other action. Both are effectively peers in the telephone system. While at first glance this might seem quite simple, it does pose a few technical challenges of which to be aware.

In application software development the concept of multiple simultaneous control interfaces is not a new one. A single application running on a computer might be controlled through its graphical user interface with inputs coming from a mouse and a keyboard, through an independent speech recognition interface, and possibly through a scripting interface such as DDE or Apple Events. The application must be prepared to combine all of theses requests into a single stream and deal with each in its turn. This is called application *factoring*. It separates from the core application code (which simply takes the next command from the outside world, processes it, and updates all the interfaces appropriately), all the code that creates and manages each of the different interfaces (different façades, if you will) and corresponding input paths. One result of application factoring is that requests that are obsolete occasionally may arrive from the interfaces. For example, if a voice command instructs a drawing application to close a file, while simultaneously a mouse gesture indicates an item is to be duplicated, the application would behave differently depending on the order in which it received the commands. If the command to close the file was received first, the command from the mouse would fail. This failure is not the result of a bug, a design flaw, or an error on anyone's part. It is just a natural consequence of an implementation in which function is factored away from the interface, and multiple interfaces are active simultaneously.

Multiple interdependent or conflicting inputs may be presented to a telephone system in a near-simultaneous fashion. To return to the first example, the computer might observe a call being presented in the *alerting* state and might react by requesting that it be deflected elsewhere. Meanwhile, however, the human sitting next to the phone (or some other computer) might already have answered it in the time it took for the computer to make its decision. This does not occur frequently in practice, but it is yet another important aspect of the CTI abstraction.

Observation and control are independent. While any request to manipulate a resource should be based on the last observed state of that resource, the request may or may not be successful because other activity may have taken place as the request was being issued. Therefore the results of a request should never be assumed; there is no substitute for observing what actually takes place.

6.1.3. Scope of Observation and Control

The portion of the telephony feature set to which a CTI interface has access varies between telephone system implementations. In telephone systems designed specifically to support CTI interfaces, the portion of functionality that can be accessed typically is much more than what can be accessed through the system's telephone sets. In systems where CTI functionality was not part of the initial design, the portion of the implemented telephony features that can be accessed through a CTI interface varies dramatically, from all to just a small subset. In some cases a system will offer both a proprietary CTI interface and a standards-based CTI interface, where the latter has less functionality than the former for historical and time-to-market reasons.

The resources making up a particular telephone system may in fact be distributed among different components, where each component may have a different view (or no view at all) of the telephony resources in the other components. Depending upon which component a particular CTI interface is associated with, its scope will be appropriately limited. Two different computers interacting with the same telephone system therefore might see different abstractions of the same system if they have access through different CTI interfaces.

6.1.4. Security

Security features can play an important role in determining what can and cannot be observed and controlled through a CTI interface.

If a particular telephone system authenticates the users of its CTI interface, the abstractions presented to two different computer systems may differ depending on their identities. A system administrator may be able to see all of the devices in a system, but another individual might only be able to see the device corresponding to his or her telephone. A secretary might be able to see the boss's phone but not have any control over it.

This type of security can be very important for avoiding the use of a CTI interface for toll fraud (see sidebar "Toll Fraud" on page 201). In practice, however, it can add a great deal of overhead to the administration and operation of the system. As a result, security features tend to be a key point of differentiation among the implementations of CTI system components from different vendors. Some customers need this type of security, but others might not.

6.1.5. Vendor Specific Extensions

No matter how extensive the abstraction, the diversity of the telephony industry dictates that individual telephone systems are likely to have features and capabilities that represent unique *vendor specific extensions*.

Most of the functionality provided and routinely used will be within the abstraction (the façade), and the majority of CTI solutions will limit their scope to just this portion of the telephone system's functionality. It is very important, however, that the CTI interface allow a "back door" or *escape mechanism* that provides direct access to these vendor specific extensions. Of course, this escape mechanism is useful only to computer systems that are aware of the precise identity of the telephone system, and are aware of precisely how its vendor specific extensions work.[6-1]

For example, the XYZ telephone system might offer the "persistent call back" feature.[6-2] This feature involves the ability to register a callback against a device so that, when the device becomes available, the telephone system calls it every 60 seconds and plays a prerecorded

---

6-1. Vendor specific extensions — One way around this limitation is using buttons. Pressing buttons on a physical device element is part of the general abstraction, but the action of a specific button is not defined in any way. Within the telephone system's abstraction, individual buttons are assigned labels to identify them and pressing the button triggers the corresponding vendor specific extension.

6-2. Vendor specific extension example — The "persistent call back feature" is purely fictitious. The author requests that no telephone system vendor try implementing this feature.

announcement requesting that the originating device be called back. The XYZ company could support this feature through the CTI interface as a vendor specific extension by using the appropriate escape mechanism. If a computer wanted to activate this feature, it would have to know it was using the right model of the XYZ system and the appropriate escape mechanism. Once activated, the persistent calling would be observed appropriately through the normal abstraction, and both the computer that set the feature and all other observers through the CTI interface would continue to accurately observe call activity.

6.2. The CTI Interface

 A *CTI interface* is a telephony resource that creates a portal through which other telephony resources can be observed by CTI components,[6-3] and through which these CTI components can request that features be set and services be carried out. CTI interfaces also may issue requests to the computer-side CTI component to perform certain tasks.

6.2.1. CTI Messages

 A CTI interface operates by generating, sending, receiving, and interpreting messages containing status information and requests for services to be performed. This is illustrated in Figure 6-3.

Messages are used in either direction both to provide information and to issue requests. The structure, content, and rules governing the flow of messages back and forth through a CTI interface is defined by a CTI protocol.

6-3. CTI Components — The word *component* is used in two different contexts in this book. A *CTI component* is a hardware or software module in a CTI system that exposes or uses a CTI interface. A *physical element component* is a lamp, button, display or some other part of a physical element.

Figure 6-3. CTI messages

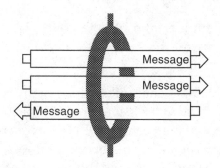

6.2.2. Parametrization

We have already seen that abstraction plays a key role in making CTI possible. In fact, with just a few simple constructs (devices, elements, components, appearances, calls, and connections) and a reasonably small vocabulary of types, states, and attribute values, we have been able to describe and model the vast majority of telephony functionality with little effort.

By translating into specific parameter values all references to parts of the abstraction and their state, status, or setting, any activity within a telephone system, desired or actual, can be described precisely.

Our abstraction of telephony resources, features, and services now can be expressed in concrete terms through parameters that are placed into messages. This is illustrated in Figure 6-4.

Figure 6-4. Parameters in a CTI message

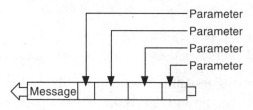

6.3. Modular CTI Systems

A general goal for implementing CTI interfaces is allowing any combination of hardware and software components to be assembled into a CTI system of any size.

Even the smallest CTI system is made up of many components, and this means that the system itself contains multiple CTI interfaces. A CTI interface is needed between each CTI component that must be integrated with another. Figure 6-5 builds on the example we viewed earlier. While this example is among the simplest of all CTI systems, it still involves three distinct components:

- The telephone

- The computer

- The CTI software running on the computer

Figure 6-5. CTI interfaces in a CTI system

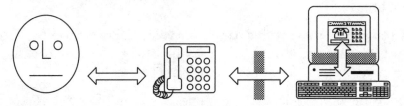

Because there are three distinct CTI components, there are two different CTI interfaces at work in this simple CTI system:

- Between the telephone and the computer:
  This CTI interface uses a protocol.

- Between the CTI software and the computer:
  This CTI interface uses a programmatic interface.

Once a CTI system is assembled, it becomes difficult to say where the telephone system begins and the computer system ends. All the components that make up the system are working together in a cohesive fashion to form what can be viewed on one hand as a more sophisticated telephone system, or on the other as a more sophisticated computer system.

Graphical Notation for CTI System Abstractions

While Figure 6-5 illustrates a tangible CTI system configuration, the rest of this chapter deals with the abstraction of CTI systems and interfaces. CTI systems and the relationships between their components are described using a standardized graphical notation. Rounded blocks represent abstract CTI components and arrows represent the flow of messages. Other symbols are introduced as they are used. These symbols are also summarized on the inside of the front cover.

The graphical notation used for describing tangible CTI system configurations is defined in Chapter 7.

6.3.1. Inter-Component Boundaries

As noted earlier, CTI systems are assembled from many individual CTI components. Regardless of a particular component's form, its role in the overall system involves exchanging CTI messages with neighboring CTI components. Each CTI component communicates CTI messages to another component through an *inter-component boundary*.

Figure 6-6. Inter-component boundary

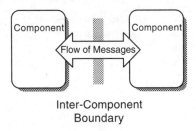

Inter-Component
Boundary

Figure 6-6 depicts an inter-component boundary through which two CTI components are exchanging messages. All the messages describing and directing CTI activity travel between two interoperable CTI system components across this boundary.

6.3.2. Logical Clients and Servers

In order to distinguish between two components exchanging CTI messages across a particular inter-component boundary one is referred to as the *logical client* and the other as the *logical server*.

The *logical server* is the CTI component, relative to a particular inter-component boundary, that is making its CTI interface available for exchanging CTI messages.

The *logical client* is the CTI component, relative to a particular inter-component boundary, that is making use of the CTI interface offered by a logical server across the inter-component boundary.

The terms *logical client* and *logical server* are always used relative to a specific inter-component boundary. They should not be confused with terms such as *client implementation*, *server implementation*, *client computer*, and *CTI server*. These terms are used as absolute references to particular types of products and are defined later in this book.

6.3.3. Organizing Components into Systems

The simplest way to organize multiple components into a system is to chain them together as shown in Figure 6-7. In this arrangement, components play the roles of both logical server and logical client (except those at either end of the chain). In this example, the second component from the left is the logical server for the component on its right and the logical client for the component on its left. The result is a "pipeline" or "bucket-brigade" in which CTI messages are passed from one component to the next.

Figure 6-7. Multi-component chain

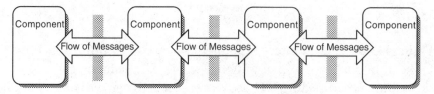

Each component in the system has some combination of the following roles:

- Generating CTI messages

- Conveying CTI messages from logical clients to logical servers

- Interpreting and responding to CTI messages

In this way components may embellish, simplify, merge, or manipulate in some fashion the CTI messages they deal with.

Figure 6-8 illustrates another arrangement of CTI components that allows for a CTI system to scale. It is referred to as a *fan-out* arrangement. A *fan-out component*, the second component from the left in the diagram, is able to simultaneously act as the logical server to multiple logical clients through a number of distinct inter-component boundaries. Each component is only aware of its counterpart across a specific inter-component boundary so the fan-out component's logical server is not aware of its logical clients and its logical clients are unaware of each other.

Figure 6-8. Fan-out component

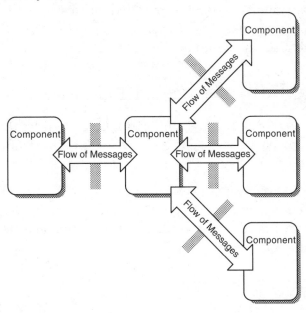

The role of the fan-out component is to maintain these relationships independently by interpreting CTI messages from each of its logical clients and either acting on them or conveying them to its own logical server. When a fan-out component conveys a CTI message in this fashion, it is effectively acting as a proxy for the logical client that generated the message in the first place. By keeping track of which CTI messages were conveyed on behalf of which clients, and in some cases by merging and separating CTI messages as they flow back and forth, the fan-out component is able to maintain all of these logical client-server relationships in a transparent fashion.

 The concept of inter-component boundaries makes possible the interoperability of actual CTI system components by standardizing only what travels between disparate components and not by standardizing their implementation.

Other benefits of this view of CTI system construction include the simplification of scaling a CTI system and the ease with which new functionality can be added to an existing CTI system. Interoperable CTI components can be added and removed from a system without affecting other components or the function of the system overall. In fact, new capabilities can be incorporated into a system by inserting CTI components with special features or capabilities into the appropriate place in a chain of other components.

6.4. Service Boundaries and Domains

In order to support a single definition for CTI interfaces that can be applied to any boundary in a CTI system of any size, configuration, or complexity, a single key insight is required:

 Regardless of the number of components in a CTI system, it can be broken down and analyzed in terms of each inter-component boundary that allows two adjacent CTI components to exchange CTI messages.

In creating or integrating different CTI components, the focus at any instant is on just one inter-component boundary through which a given component must interoperate. CTI messages and the CTI interfaces that work with these messages can therefore be described in terms of a simplified CTI system which consists of just three parts:

- Switching domain

  The *switching domain* is everything on the logical server's side of a particular inter-component boundary.

- Computing domain

  The *computing domain* is everything on the logical client's side of a particular inter-component boundary.

- Service boundary

  The *service boundary* is the inter-component boundary that lies between the computing domain and the switching domain in a given context.

This simplification of a CTI system is illustrated in Figure 6-9. All interaction between a given computing domain and a switching domain takes place through a CTI service boundary.

Figure 6-9. The CTI service boundary

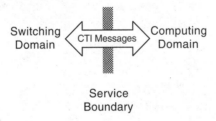

This abstraction applies to every inter-component boundary in a CTI system so care must be taken to specify what inter-component boundary is being referred to as a service boundary at any given instant. Specifying the service boundary not only identifies which pair of components is being discussed, but also the CTI messages that travel between them, and the CTI interface that interprets these messages.

6.4.1. CTI Service Boundary

A *CTI service boundary* is the inter-component boundary through which CTI messages from a CTI component acting on behalf of its computing domain are communicated to the CTI interface of a CTI component acting on behalf of the switching domain it represents.

In concrete terms, a service boundary can take the form of either a protocol or a programmatic interface. A service boundary that lies between two software components running on the same hardware component (e.g., a computer) can take the form of a *programmatic interface* through which CTI messages are passed using function calls. Otherwise, the service boundary takes the form of a *CTI protocol* used for conveying CTI messages as a stream of data.

6.4.2. Switching Domain

The term *switching domain* refers to all of the telephony resources that can be observed or controlled through a designated service boundary and all of the CTI components that provide this access.

The mechanism a switching domain uses to interact with a computing domain through the service boundary is its CTI interface. The switching domain encompasses any telephony resources associated with components in the switching domain that can be accessed using this CTI interface. This is illustrated in Figure 6-10.

Figure 6-10. The switching domain

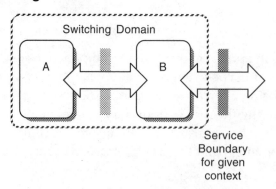

340

In concrete terms, the number and type of hardware and software components that might be found in a given switching domain is unbounded. However, at least one component must be among the telephony products presented in Chapter 5. These include:

- Switches
    - Front-end switches
    - KSUs or Hybrids
    - PBXs
    - Application specific switches
    - Internet voice gateways

- Telephone station equipment
    - Telephone station
    - Telephone station peripherals

Any type of component that can be found in a CTI system may be found in a switching domain given the service boundary context. Other types of CTI hardware components are presented in Chapter 7 and software components are presented in Chapter 8.

6.4.3. Computing Domain

The term *computing domain* (as illustrated in Figure 6-11) refers to all of the CTI components that are involved in observing or controlling telephony resources in the switching domain on the other side of a designated service boundary.

Figure 6-11. The computing domain

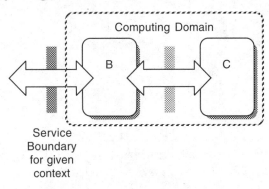

341

Any type of CTI component may be found in a computing domain, given a particular service boundary context. However, at least one component is a software component that is attempting to observe or control telephony resources in the switching domain. (See Chapter 7 for details on CTI hardware components and Chapter 8 for details on CTI software components.)

6.4.4. Service Boundary Context

The abstraction of switching domain, computing domain, and service boundary can be applied to any inter-component boundary in a CTI system of any size. For a given context, the applicable service boundary always determines the roles of the components on either side of the boundary by whether they fall within the computing domain or the switching domain.

With respect to achieving interoperability between the two domains at a given point, the central concept is the service boundary: an inter-component boundary between the two domains over which well-defined CTI control and status messages pass.

 Regardless of the number of components in a given CTI system, each can be viewed as being interfaced to its neighbor through a particular service boundary.[6-4] This is illustrated in Figure 6-12.

Figure 6-12. Service boundary contexts

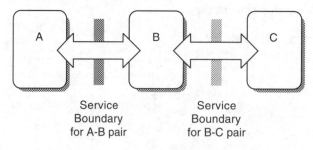

Service Boundary for A-B pair

Service Boundary for B-C pair

6-4. CTI components and service boundaries — In a CTI system consisting of *n* interoperable components, there are *n*-1 service boundaries.

As shown in Figure 6-13, the particular service boundary for a given context can be established by referencing the pair of components which share it. When any given pair of CTI components are referenced (such as components "B" and "C" in the example), the implied boundary is considered the service boundary, with the computing domain on one side and the switching domain on the other, forming a domain pair.

Figure 6-13. Service boundary defines switching domain and computing domain

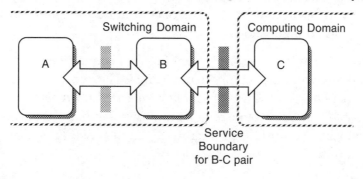

In fact, given a particular division of components between the switching domain and the computing domain, the service boundary is unique. There is exactly one service boundary over which messages pass between any computing domain-switching domain pair.

Applying these concepts, an unlimited variety of CTI configurations can be supported; from the simple interconnection of two CTI products, to a whole system of interconnected CTI components. CTI control and status messages are communicated over these service boundaries, and each component may add value to the services provided by its neighbor or simply act as a conduit. The CTI system as a whole is able to function because of the individual service boundaries that hold its components together. (See Chapter 7 for examples of CTI system configurations made possible by this modularity.)

6.4.5. Protocols

 CTI protocols are specifications of the structure, contents, use, and flow of CTI control and status messages that travel between CTI system components over well-defined communication paths. CTI protocols are high-level protocols, like the protocols used to send electronic mail, print to a printer, retrieve files from a file server or, browse the World Wide Web. Like these other protocols, they are designed to be transmitted over any type of reliable communication path.[6-5] CTI protocols are applicable to all types of communication paths and to all types of CTI configurations.

As shown in Figure 6-14, implementations of components that interoperate with other components using CTI protocols include a subcomponent referred to as a *CTI protocol encoder/decoder*, which is responsible both for establishing communication paths that carry the CTI protocol, and for interpreting the CTI protocol that flows across the communication path.

Figure 6-14. CTI protocols

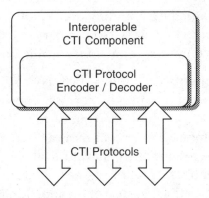

6-5. CTI Protocols, OSI layers — In the terminology of Open Systems Interconnection (OSI) layering, CTI protocols are layer 7 protocols with defined layer 6 encodings. They do not assume any particular underlying protocol stack. Communication paths provide OSI layers 1–5 in the form of functionally acceptable session/transport protocol stacks.

6.4.6. Programmatic Interfaces

Programmatic interfaces are mechanisms, typically made up of function calls, that allow two software components present on the same hardware component (generally a computer of some sort) to link to one another and exchange messages.

There are three distinct categories of programmatic interfaces of concern to CTI systems.

- Read/Write interfaces

> R/W Interface

Read/write interfaces, or just *R/W interfaces* for short, are simple interfaces that allow a software component to obtain access to a communication path carrying a CTI protocol stream.

- Procedural interfaces

Procedural Interface

Procedural interfaces allow a software component to obtain CTI functionality through a set of procedural function calls.

- Object interfaces

Object interfaces allow a software component to obtain CTI functionality by manipulating software objects.

The use of programmatic interfaces for CTI is described extensively in Chapter 8.

6.4.7. Fundamental CTI System Configurations

To illustrate the concepts of switching domains, computing domains, service boundaries, protocols, and programmatic interfaces, we'll look briefly at two fundamental examples of actual CTI system configurations. The following sections show examples of basic *direct-connect* and *client-server* CTI system configurations.[6-6] (A detailed discussion of these and other CTI configurations is presented in Chapter 7.)

CTI interfaces, accessed through CTI service boundaries, allow the integration of CTI system components with varying functionality and from different vendors (such as the client computers, telephone stations, CTI servers, and switches shown in the examples below).

Direct-Connect Configuration

This first example, depicted in Figure 6-15, involves a simple direct-connect[6-7] CTI system configuration. Two CTI hardware components are connected with a serial cable. One is a personal computer running a CTI application and the other is a telephone station.

There are two service boundaries in this configuration. The first is a protocol boundary that lies between the telephone station and the personal computer. The second boundary is a programmatic interface within the personal computer that supports the CTI application.

Figure 6-15. Service boundaries in a direct-connect configuration

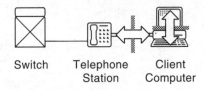

Switch   Telephone   Client
Station   Computer

Figure 6-16. Direct-connect example CTI components

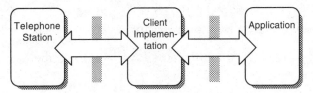

Telephone Station — Client Implemen-tation — Application

6-6. Configurations versus call control models — The terms *direct-connect* and *client-server* (which describe configurations) should never be confused with the concepts of first-party and third-party call control. (The terms *first-party* and *third-party* are explained in sections 6.5.2 and 6.5.3.)

6-7. Direct-connect configuration — A *direct-connect* configuration involves direct interconnection of telephone station equipment with computer equipment. Communication paths generally take the form of serial cables/buses or add-in cards. See Chapter 7 for details on CTI system configurations.

Figure 6-16 shows the same configuration in terms of abstract CTI components. The telephony resources in this example are all ultimately accessed through a CTI interface associated with the telephone station. The telephone station is a CTI hardware component that uses telephone resources in the switch, however, the switch is not considered a distinct component of the CTI system because no CTI interface exists between the telephone station and the phone and thus no boundary is present.

The telephone station exchanges CTI messages with the personal computer using a protocol. Within the personal computer, it is actually a software component known as a *CTI client implementation*,[6-8] that is interacting with the telephone station. It interprets the protocol from the telephone station and, in turn, directs CTI messages to the application (and vice versa) using the programmatic interface.

From the telephone station's perspective, the CTI client implementation is in the computing domain; from the application's perspective, however, the CTI client implementation is part of the switching domain.

Client-Server Configuration

This second example (Figure 6-17), involves a client-server[6-9] CTI system configuration. This configuration involves one more component than in the direct-connect example so there is one additional boundary in the system. A personal computer running a CTI application is connected to a CTI server hardware component over a local area network. The CTI server is, in turn, connected to a switch.

6-8. CTI client implementation — CTI client implementations are CTI software components in the CTI software framework that is presented in Chapter 8.
6-9. Client-server configuration — A *client-server* configuration involves indirect control of telephony resources using a fan-out component in the form of a CTI server. The CTI server sits between client computers and the telephony resources being accessed. Despite the fact that there is no direct connection between the client computer and the telephone station being manipulated in this configuration, logical integration takes place through the indirect path of CTI messages. See Chapter 7 for details on CTI system configurations.

There are three service boundaries in this configuration. The first is a protocol boundary between the switch and the CTI server. The second is another protocol boundary between the CTI server and the personal computer. The third is a programmatic boundary used by the application software.

Figure 6-17. Service boundaries in a client-server configuration

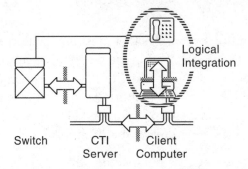

Switch   CTI   Client
         Server   Computer

Figure 6-18. Client-server example CTI components

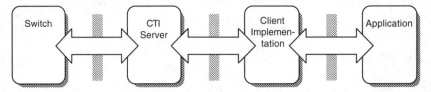

Figure 6-18 shows this example in the form of abstract CTI components. Each service boundary represents a different pair of switching domain-computing domain combinations. The one being discussed at any given instant depends upon the context. If the service boundary in question is the programmatic interface between the application and the client implementation components, the switching domain implementation includes the switch, the CTI server, and the client implementation. On the other hand, if the service boundary in question is the cable between the CTI server and the switch, the switching domain consists only of the switch.

6.5. Switching Domain Abstraction

The emphasis in this chapter is primarily on the CTI functionality that the switching domain offers to the computing domain. (Chapter 8 will look more closely at the implementation and functionality of the computing domain.)

From the perspective of the computing domain, the switching domain is the CTI system that is being observed or controlled. The computing domain is completely unaware of the actual implementation of the telephony resources in question, or of the physical topology and components making up the CTI system. The reality of the implementation and the abstraction of the switching domain may be closely related or very different—but this is completely irrelevant as far as the computing domain is concerned. The only telephony resources, features, and services of which the computing domain is aware are the ones that can be observed in the switching domain.

Throughout the remainder of this book, the term *switching domain* will be used to refer to the abstraction, or version, of the CTI system that can be observed through CTI technology, and the term *CTI system* will be used in reference to the configuration and tangible components of the CTI system.

6.5.1. Switching Domain Scope

Switching domain *scope* refers to the set of telephony resources in a particular switching domain—devices and calls in particular—that can be observed or controlled. External networks represent other sets of telephony resources outside a given switching domain. The switching domain in question may or may not have direct access to a particular external network and its resources. This is illustrated in Figure 6-19.

Most resources inside a switching domain are visible, that is, they are directly observable and are within the scope of the switching domain. Certain telephony resources, however, may be invisible. This means that their presence may be known and indirectly used, but they cannot be directly referenced or observed. Observation of connections is

Figure 6-19. Switching domain scope

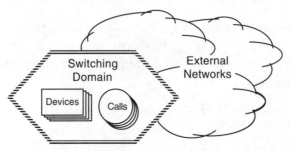

assured only within the scope of the switching domain (i.e., for visible devices inside the switching domain). The status of other connections cannot be relied upon as they may or may not be reported, depending on the implementation of the switching domain and the nature of the external network.

Switching domain scope is a very important concept because it indicates what can and cannot be accomplished using a given CTI interface. For example, if a call is placed to a device outside the switching domain, the computing domain cannot obtain any information about the status (or even existence) of the called device, and it cannot be assured of connection state information for the called device.

6.5.2. First-Party Call Control

First-party call control is a call control model in which only a single device or device configuration can be observed and controlled. If a particular CTI interface supports first-party call control, the scope of the associated switching domain contains only a single device or device configuration. This is illustrated in Figure 6-20.

Because the switching domain has only one visible device configuration, all calls are either external incoming or external outgoing.

Figure 6-20. First-party call control

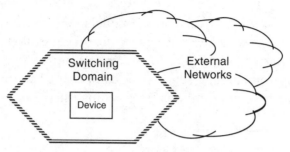

Though first-party call control may appear quite limiting with respect to the many functions actually being performed within a telephone system, this level of functionality is all that is required for the majority of CTI applications.

The CTI system configuration has no bearing on what telephony resources are in a given switching domain or vice-versa. The CTI interface presenting the single-device (i.e., first-party) switching domain may be an individual telephone, a CTI server, a switch, or any other CTI component in a given system configuration. Figure 6-21 shows a CTI system configuration involving a CTI server that is providing the computing domain with visibility of a single device only. (Refer to Chapter 7 for more information on CTI system configurations.)

Figure 6-21. First-party call control in a CTI system

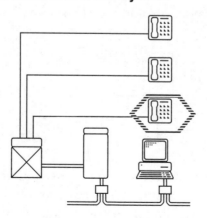

6.5.3. Third-Party Call Control

Third-party call control is a call control model in which multiple devices, or device configurations, can be observed and controlled simultaneously. If a particular CTI interface supports third-party call control, the switching domain it presents is one comprising two or more visible devices or device configurations. This is illustrated in Figure 6-22.

Figure 6-22. Third-party call control

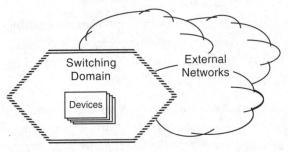

As noted earlier, the CTI system configuration has no bearing on what telephony resources are in a given switching domain. Figure 6-23 shows a CTI system configuration involving a CTI interface exposed by an individual telephone station. Despite this, the switching domain presented contains multiple devices and is thus an example of third-party call control. (Refer to Chapter 7 for more information on CTI system configurations.)

Figure 6-23. Third-party call control in a CTI system

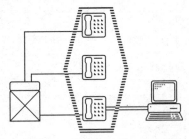

Comparing Figures 6-20 and 6-21 to Figures 6-22 and 6-23, it is clear that the only difference between first-party and third-party call control is the scope of the switching domains presented by the CTI interface implementations. Third-party call control is effectively the general case, and first-party call control is a special case in which only one device is visible.

6.5.4. External Network

Any device or other telephony resource not within the switching domain is, by definition, part of an external network. Network interface devices are special devices that exist in both the switching domain and one or more accessible external networks. Network interface devices may or may not be visible within a switching domain. Because network interface devices exist simultaneously inside and outside the switching domain, they are used as proxies. (Refer to Chapter 3, section 3.4.5 for more information on network interface devices as proxies.)

If a given network interface device is invisible, so that it cannot be observed directly, information about the state of connections to the network interface device may or may not be provided.

6.6. CTI Service Requests and Events

There are four kinds of messages[6-10] that pass through a CTI interface:

- Events

- Service requests

- Positive acknowledgments

- Negative acknowledgments

6-10. CTI messages — CTI messages are implemented within CTI protocols as self-contained *protocol data units* (PDUs). On the other hand, procedural and object-based programmatic interfaces translate messages to sequences of functions, parameters, function return codes, call back routines, and data structures. In the remainder of this chapter, CTI messages are presented in the context of the standard CTI Plug & Play protocols. (See Chapter 7 for more information about standard and proprietary CTI protocols, and Chapter 8 for a discussion of standard CTI programmatic interfaces.)

Every CTI message is defined in terms of its kind, which message among those of its kind it is, and its parameters. For example, the CTI message corresponding to a request for the *set lamp mode* service can be shown graphically as in Figure 6-24.

Figure 6-24. Set Lamp Mode service request message

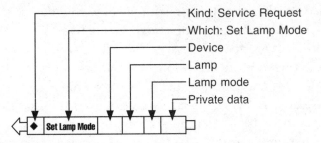

Kind: Service Request
Which: Set Lamp Mode
Device
Lamp
Lamp mode
Private data

Each parameter in the list of parameters appropriate for a given message may be either, optional, mandatory, or conditionally mandatory.[6-11] Most parameters are either identifiers that reference a particular resource (or resource attribute) in the switching domain, or they are variables representing a state, status, or setting value.

6.6.1. CTI events

CTI event messages, or just *events* for short, are messages sent from the switching domain to the computing domain to indicate transitions of states and changes in the status or setting of an attribute in the switching domain. Events are the primary mechanism used by the computing domain to observe activity within the switching domain.

For example, if the connection state of a particular connection transitions from *alerting* to *connected*, the CTI event message *established* would be sent to the computing domain to indicate that the connection

---

6-11. Parameter optionality — The following annotations are typically used to indicate the requirements for a particular parameter: "M" means *mandatory*, "O" means *optional*, and "M/O" means *conditionally mandatory*.

in question had transitioned to the *connected* state. The established event (with a partial parameter list) is shown graphically in Figure 6-25.

Figure 6-25. Established event message

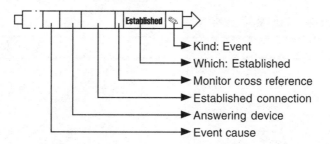

In order to receive event messages that are relevant to a particular call or device in the switching domain, the computing domain must request them by *starting a monitor* on the item in question. (Monitoring is described in section 6.9.3 of this chapter.) The monitor cross-reference identifier parameter in the event identifies what previously established monitor caused this event to be sent.

The definition of event messages include:

- The meaning of the event and, in the case of events that reflect state changes, its context in terms of implications for other related connections;

- Mandatory, optional, and conditionally mandatory parameters; and

- The possible causes to which the event can be attributed. (The *cause code* of a given event is an essential parameter in the event message, and in many instances it represents a very important clarification of the meaning of an event.)

Event messages are defined for every type of item that has a state, status, or setting that can change. Events also are defined to indicate that new information has been received, such as an update to the correlator information associated with a call or the detection of a DTMF tone. (Refer to Table 6-1 for more examples of events.)

Table 6-1. Event message examples

Event messages	Resource	Event Indicates
Call Cleared	Call	Call no longer exists
Call Information	Call	Updated call associated information
Bridged	Connection	Transition to *queued* state (during shared bridging)
Connection Cleared	Connection	Transition to *null* state
Delivered	Connection	Transition to *alerting* state
Digits Dialed	Connection	Transition to *initiated* state (digits were dialed)
Established	Connection	Transition to *connected* state
Failed	Connection	Transition to *fail* state
Held	Connection	Transition to *hold* state
Offered	Connection	Transition to *alerting* state (*offered* mode)
Originated	Connection	Transition to *connected* state (after originating a call)
Queued	Connection	Transition to *queued* state
Retrieved	Connection	Transition to *connected* state (after retrieve)
Service Initiated	Connection	Transition to *initiated* state
DTMF Digits Detected	Connection	DTMF digits detected
Telephony Tones Detected	Connection	Telephony tones detected
Button Press	Button Component	Button was pressed
Display Updated	Display Component	Updated display contents
Hookswitch	Hookswitch Component	Change in hookswitch status
Lamp Mode	Lamp Component	Change in lamp mode
Microphone Mute	Microphone Component	Microphone mute attribute updated

Table 6-1. Event message examples (Continued)

Event messages	Resource	Event Indicates
Ringer Status	Ringer Component	Change in ringer attribute
Speaker Volume	Speaker Component	Speaker volume attribute updated
Agent Logged On	Agent	Transition to *agent logged on* status
Agent Not Ready	Agent	Transition to *agent not ready* status
Do Not Disturb	Logical Element	Do not disturb setting changed
Forwarding	Logical Element	Forwarding settings changed
Out of Service	Device Configuration	Device is out of service

The example in Figure 6-26 illustrates how event sequences communicate what is taking place in the switching domain. This example revisits the scenario involving parking a call, which was presented in Figure 4-23 of Chapter 4, section 4.7.6. The scenario involved an attendant, D2, parking a call to park device D3 and then having it picked up by device D4. Figure 6-26 shows what's happening to the devices and connections in the switching domain on the left, and the corresponding event sequence generated by a monitor started for device D1 on the right. It is important to note in this example that the event sequence shown is just the sequence for D1. If the other devices and/or the call itself were being monitored, each corresponding monitor would generate a similar sequence of events. In this case, *diverted* event messages indicate that the connection indicated in the event has been cleared, and that the call is about to be associated with the new destination device shown. The *queued* event indicates that the connection D3C1 is in the *queued* state and has transitioned to this state because it was parked. Finally, the *established* event indicates that the connection D4C1 is in the *connected* state because it was picked.

6.6.2. Service Requests

Service requests are messages sent by either domain to request some service of the other. The vast majority of services are *switching domain service requests*. These correspond to services that the switching

Figure 6-26. Park and pick scenario example event flow

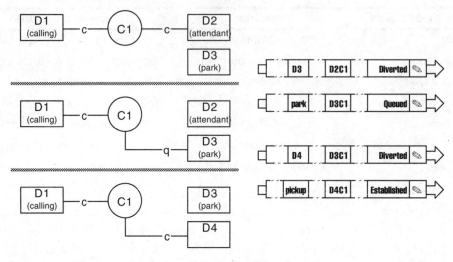

domain can carry out. *Computing domain service requests* are messages sent to the computing domain to request that it perform some function. *Bidirectional service requests* can be sent in either direction.

There are three categories of service requests:

1. Service requests associated with the telephony features and services explored in Chapter 4. These include:

   - Call control services

   - Call associated services

   - Logical device services

2. Service requests that manipulate or check the status of physical element components. These services include:

   - Pushing buttons

   - Getting and setting button information

   - Getting lamp information

   - Getting and setting lamp mode

   - Getting and setting display contents

- Getting and setting message waiting indicator

- Getting auditory apparatus information

- Getting and setting hookswitch status

- Getting and setting microphone gain and mute

- Getting and setting speaker volume and mute

- Getting and setting ringer status

3. Service requests that are specific to the CTI interface. These services will be discussed at greater length through the rest of this chapter. These include:

- Capabilities exchange

- System status services

- Monitoring

- Snapshot services

- Routing

- Media access

- Vendor specific extensions

Service Request Messages

The definition of service request messages include:

- The service that is invoked or the feature that is set;

- Required initial states and possible final states for any connections on which the service acts;

- Mandatory, optional, and conditionally mandatory parameters of the service request;

- Possible outcomes of the service;

- The sequence of events that should be expected if the service completes successfully;

- The completion criteria used to determine if a service completed successfully; and

- The possible reasons for an unsuccessful service request.

When a service request message is issued by one domain, the other domain responds[6-12] with a positive or negative acknowledgment. Independent of these acknowledgments, if the service results in any action that affects the state of one or more connections, or the status or setting of some resource or attribute, the switching domain will generate all the appropriate event messages.

The event sequence defined as part of the service completion criteria for each service is referred to as a *normalized event flow*, or just *flow* for short. It is very important because it allows the computing domain to verify that a service has taken place. As noted earlier, a given computing domain is not the only source of commands manipulating items within the switching domain. If another interface, such as a telephone set, is used to issue a command, the computing domain sees the results of this command as an event sequence. The event sequence observed allows it to determine what just took place. For example, the event sequence for the completion of a *consultation call* service (with single-step dialing) is shown in Figure 6-27. (Refer to Chapter 4, section 4.11.2 for a description of the *consultation call* service.) In this case the first step, placing the connection to the initial call in the *hold* state, is indicated with the *held* event. The fact that a new call is created for the consultation call is indicated through the *service initiated* event.[6-13] Finally, the *originated* event indicates that the second call has been originated and connection D1C2 is in the *connected* state. Note

---

6-12. Service request responses — There are a small set of service requests for which no response messages (positive or negative acknowledgments) are required (or defined).

6-13. Service initiated — The *service initiated* event in this flow is optional. If no prompting of the device is required, the newly created connection could go directly to the *connected* state because single-step dialing was being used (i.e., the complete called device address was provided as a parameter to the consultation call service).

that while the *consultation call* service specifies that the final state of connection D1C2 must be one of those shown, the completion criteria for the service are satisfied when the *originated* event is provided.

Figure 6-27. Consultation Call event sequence example

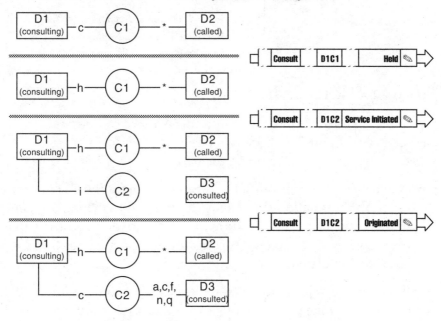

6.6.3. Negative Acknowledgments

If a service request is unsuccessful for some reason, the domain attempting to carry out the service indicates the lack of success by sending a negative acknowledgment message. This negative acknowledgment message contains an error value that provides an explanation as to what the problem was. An example of a negatively acknowledged service request is illustrated in Figure 6-28.

Figure 6-28. Negative acknowledgment sequence example

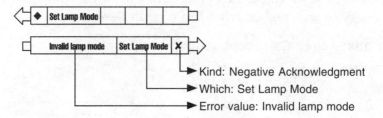

6.6.4. Positive Acknowledgments

A positive acknowledgment indicates that a service request is being or has been acted upon. There are two basic types of positive acknowledgments, depending on the nature of the service request:

- One type of service request is a direct request for the CTI interface to return information about something in the switching domain. In this case, the positive acknowledgment not only indicates that the request was completed successfully, it also includes the requested information.

- If the service request involved requesting that some manipulation of resources take place, the positive acknowledgment indicates that the CTI interface has passed the request to call processing to be carried out. If the service in question is carried out in an atomic fashion, the positive acknowledgment also indicates notification that the service was completed successfully.

An example of a positively acknowledged service request is illustrated in Figure 6-29. The service request in this example is a request for information.

The definition of a positive acknowledgment message for a given service may include mandatory, optional, and conditionally mandatory parameters.

Figure 6-29. Positive acknowledgment sequence example

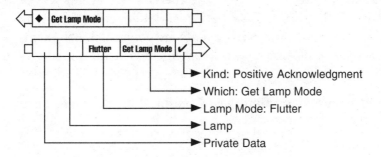

6.6.5. Atomic and Multi-Step services

The implementation of services in the switching domain may be atomic or multi-step. The nature of the implementation determines the correct interpretation of positive and negative acknowledgment messages.

If a particular service is implemented in an atomic fashion, the switching domain treats requests for it as follows:

1. Validation

    Are all of the parameters valid? Are all applicable connections in the correct states? Are all necessary resources available? If the service request is invalid for any reason, the switching domain sends back a negative acknowledgment message to indicate that the request has failed.

2. Execution

    The switching domain then attempts to carry out the service requested.

3. Acknowledgment of success or failure

    If the service request succeeded, the switching domain sends a positive acknowledgment message to indicate it has succeeded. Otherwise it sends a negative acknowledgment, indicating that the service did not succeed and why it did not.

If an atomic service does not succeed, no resources in the switching domain are affected in any way and therefore no events will be generated. If the service is successful and it affects states, statuses, settings, etc., then appropriate events are generated. For example, if the *consultation call* service is implemented as an atomic service, the complete flow might be as shown in Figure 6-30.

Figure 6-30. Atomic implementation of Consultation Call service

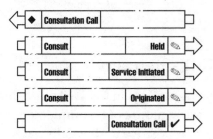

If a service is implemented in a multi-step fashion, the request is processed somewhat differently:

1. Validation

    As in the atomic implementation, all of the parameters are first validated, initial states are checked, and the availability of all required resources is verified. In the multi-step implementation, the decision to attempt the service or to reject the service request is made at this point. If the validation does not succeed, the switching domain returns a negative acknowledgment. In a multi-step implementation, this is the only point at which a negative acknowledgment will be generated. If everything appears to be in order, the switching domain returns a positive acknowledgment. However, this positive acknowledgment indicates only that the request has been passed to call processing to be executed and that there is every expectation of success; it does not indicate that any execution has begun or that it has satisfied the completion criteria.

2. Execution

The switching domain begins to carry out the multi-step service after the positive acknowledgment message is sent. As the execution of the service proceeds and connection states are affected, statuses are changed, etc., the switching domain generates appropriate events. The computing domain uses these events to determine when the service has completed by comparing the events received to those stipulated in the definition of the service request. Assuming that the service does succeed, there will be no other indication of success.

3. Reporting incomplete service execution

If the service fails to succeed for any reason, the switching domain sends a special event message known as a *service completion failure* event to indicate that the execution of the service could not be completed.

This event is not related to the *fail* state in any way. It merely indicates that one or more of the completion criteria associated with a particular service could not be satisfied.

The multi-step implementation of *consultation call* is illustrated for both the case of success and the case of no success in Figures 6-31 and 6-32 respectively.

Figure 6-31. Multi-step implementation of Consultation Call service (succeeds)

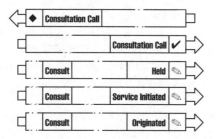

The switching domain uses capability exchange services (discussed in section 6.8.2) in order to determine which services a given switching domain implements as atomic and which it implements as multi-step.

Figure 6-32. Multi-step implementation of Consultation Call service (does not succeed)

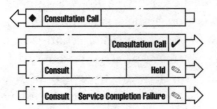

6.7. Identifiers: Referencing Switching Domain Objects

Identifiers are message parameters used to identify a particular resource or entity in the switching domain. They allow the abstraction represented by the switching domain to be conveyed across the service boundary via messages.

6.7.1. Device Identifiers

Device identifiers are parameters that refer to devices and device configurations. Devices are referred to by address, so device identifier parameters contain device addresses in a *device identifier format* corresponding to one of the address formats described in Chapter 3, section 3.10.

Device Roles

A particular service request message or event message may contain many references to different devices. Each of these devices has a different role with respect to the service desired or the event being reported.

These *device roles* include:

- Called device identifier

   The *called device identifier* is the destination address specified for a particular call. For an external incoming call, this parameter will contain DNIS information or the full DID number dialed.

- Calling device identifier

   The *calling device identifier* is the device that originally placed a particular call. For an external incoming call, this parameter will contain callerID or ANI information.

- Associated called device identifier

   For an external outgoing call, the *associated called device identifier* parameter refers to the network interface device being used. For an external incoming call, the associated called device identifier parameter refers to the device in the switching domain associated with the number originally called (e.g., the attendant, DID extension, DISA extension, etc.). If the call is not an external call, this parameter is not included in any messages.

- Associated calling device identifier

   For an external incoming call, the *associated calling device identifier* parameter refers to the network interface device being used. If the call is not an external call, this parameter is not included in any messages.

- Redirection device identifier

   The *redirection device identifier* parameter refers to the last device from which this call was previously routed. In addition to being Not Known (which means that the last redirection device is not visible), the redirection device identifier may be specified as Not Required to indicate that the call has not been redirected, or Not Specified to indicate that the switching domain doesn't know if the call has been redirected.

- Subject device identifier

  The *subject device identifier* parameter indicates the subject, or focus, of a particular event message.

Depending on the context, a particular device that must be referenced in an event message might not be visible within the switching domain. Parameters referencing these devices have the value *Not Known*.

Addressable and Non–addressable Appearances

Specific appearances within a logical device element may be referenced explicitly if the appearances are addressable. (Addressability of appearances is explained in Chapter 3, section 3.8.2.) A device identifier containing explicit appearance references is formed using appearance suffixes. (See the sidebar "Switching Domain Representation Format" on page 165 in Chapter 3, section 3.10.5.)

Physical, Logical, Appearance, and Device Configuration References

The interpretation of a device identifier parameter depends upon the value provided and the service request or event in which it appears.

In the case that a device identifier contains an appearance suffix (see Chapter 3, section 3.10.5), the context in which it is being used determines its interpretation. If the context is one where an appearance reference is appropriate, it is used for this purpose; otherwise the parameter is treated as a reference to the appearance's logical device element.

If the device identifier is used in a context in which only a physical element reference is applicable, it is interpreted as being a reference to the physical element portion of the device specified.

Likewise, if the device identifier is used in a context in which only a logical element is applicable, the device identifier is interpreted as referring to the logical element portion of the device specified.

Otherwise, the device reference is interpreted as being the device configuration which has the device referenced as its base.

6.7.2. Physical Element Component Identifiers

References to the components of physical elements are made by specifying a particular physical device element using an appropriate device identifier and providing the identifier for the desired component. The component's identifier may be one of the following:

- Hookswitch identifier

- Auditory apparatus identifier

- Button identifier

- Lamp identifier

- Ringer identifier

A reference to an auditory apparatus identifier is used whenever referring to a particular auditory apparatus, or when referring to either the speaker or microphone associated with that auditory apparatus.

6.7.3. Call and Connection Identifiers

Calls are referenced using parameters called *call identifiers*. They are numbers that uniquely identify individual calls in the switching domain.

Connection identifiers uniquely reference a particular connection in the switching domain. Because a connection represents the association of a call and a device, a connection identifier is formed simply by combining a device identifier and a call identifier. If the device is one that is not visible to the switching domain, or the connection involves a non-addressable appearance, the switching domain will form the connection identifier using a private or *dynamic device identifier*.[6-14]

6-14. Connection Identifiers — The device identifiers contained within connection identifiers are not to be extracted by the computing domain because they may or may not be valid outside the context of the connection identifier.

Call and Connection Roles

Services that involve multiple calls involve both service request and event messages that include multiple connection identifiers. In these cases (typically transfer- and conference-related services), the different calls referenced have different roles with respect to the service. These roles include:

- Primary call

     The *primary call* is the first of two calls being operated upon. It is the call placed on hold by a *consultation call* service request when setting up for a transfer or conference service.

- Secondary call

     The *secondary call* is the second of two calls being operated on. It is the new call created by a *consultation call* service request when setting up for a transfer or conference service.

- Primary old call

     The *primary old call* is a reference to the primary call that was merged into a new call as a result of a switching service involving the merging of two calls.

- Secondary old call

     The *secondary old call* is a reference to the secondary call that was merged into a new call as a result of a switching service involving the merging of two calls.

- Resulting call

     The *resulting call* is a reference to the new call that is the result of a switching service involving the merging of two calls.

6.8. CTI Interoperability

Building a CTI system involves assembling many independent CTI components by attaching each one to another through an appropriate service boundary. As CTI technology becomes increasingly ubiquitous, CTI components (hardware and software) will become increasingly interoperable through the use of standard interfaces: CTI protocols for (hardware and software components), and programmatic interfaces (for software components).

A measure of increasing CTI component interoperability is the level of human intervention involved in getting them operational. Market forces will see to it that the use of standard CTI protocols assures system integrators, customers, and individuals of getting full benefit from the potential that CTI Plug & Play has to offer. Human involvement in connecting two CTI components should be limited to ensuring that the appropriate physical communication path is in place and instructing one component where to find the other. The rest of the information needed for the two components to interoperate should be determined dynamically by the two components through negotiation and exchange of capability information.

6.8.1. Protocol and Version Negotiation

Before two components in a CTI system can begin exchanging messages across their service boundary, they first must determine what protocol and/or CTI interface version is to be used. In addition, if the switching domain implementation supports vendor specific extensions, these have their own *private data version* that is independent of the version of the CTI interface used.

When two CTI Plug & Play hardware components first establish a communication path, they exchange *protocol negotiation packets*. The computing domain first sends the switching domain a protocol negotiation packet that indicates the range of protocols and versions supported, and an indication as to whether or not private data negotiation is desired. The switching domain replies by choosing the

371

protocol and version it wishes to use (or indicating that there is no version that is supported in common, so the two components can gracefully disconnect). If private data negotiation was requested, the switching domain provides the information necessary for this as part of its response, and the computing domain responds with a service request indicating the private data version it has chosen. The negotiation process completes when the switching domain indicates its system status (see section 6.9.1) and the computing domain begins the process of learning about the switching domain using the capabilities exchange services (described in the next section).

Similar mechanisms are implemented in the programmatic CTI interfaces (APIs) used between CTI software components. In addition to being used by CTI client implementations and applications, these mechanisms are used by API-specific adapter software associated with CTI components that don't support standard CTI protocols directly or through mappers.

6.8.2. Capabilities Exchange

Once two components have agreed on a CTI version to use, the computing domain must learn about the switching domain before it can begin to observe or manipulate resources within it. The first thing the computing domain must do is find out what the general capabilities of the switching domain are. This includes finding out what capabilities exchange services are supported.

Switching domain implementations that support CTI Plug & Play implement the *get switching domain capabilities* service. This service is a request for information; it returns the desired capabilities information in the positive acknowledgment to the service request. The positive acknowledgment reports such things as:

- The name of the switching domain implementation. This name identifies the vendor and model of the switching domain being used. This name is used for determining what, if any, vendor specific extensions might be applicable.

- A default or suggested device within the switching domain that the switching domain believes is associated with the computing domain. If provided, this saves the computing domain from having to ask a human user what device it should monitor and manipulate by default.

- Which device identifier formats are supported.

- Whether the switching domain supports external incoming or external outgoing calls.

- If and how the forwarding feature is supported.

- If *dynamic feature availability* (see next section) is supported.

- The time indicated by the switching domain's internal clock.

- Maximum sizes of certain variable-sized parameters.

- What services and events are supported and of these what optional parameters are supported.

Once this information has been digested, the computing domain then typically tries to find out what devices are visible in the switching domain. The *get switching domain devices* service allows it to find all the visible devices in the switching domain and determine the type and device identifier for each.

Once the computing domain has determined what device(s) it is interested in observing and/or controlling, it uses the *get logical device information* and *get physical device information* services to learn about each device of interest.

The *get physical device information* service provides such information as:

- Product name of the physical device

- Size of the display (if present)

- Number of buttons (if any)

- Number of lamps (if any)

- Number of ring patterns supported

- Physical element-related services and events supported by a given physical element, and the optional parameters supported for each

- Logical elements associated through the device configuration for which the given physical element is the base

The *get logical device information* service provides information that includes the:

- Maximum number of callback requests supported for the device

- Maximum auto-answer value supported

- Maximum number of connections supported

- Maximum number of held calls supported

- Maximum number of forwarding rules supported

- Maximum number of devices that can be conferenced together

- Media services supported by the logical device or available in conjunction with the logical device

- Supported ways to prepare to transfer or conference a call

- Logical element related services and events supported by a given logical element, the optional parameters supported for each, and the required initial states associated with call control services

- Physical elements associated through the device configuration for which the given logical element is the base

- Type, behavior, and addressability of the logical element's appearances

A key step in the computing domain's preparation sequence for working with the switching domain involves figuring out what services are supported and, for services involving call control, when these services may be requested for a given device. The information provided by the *get logical device information* service, relative to the services supported and associated initial states, is used to build up this picture.

6.8.3. Dynamic Feature Availability

Dynamic feature availability is a feature that may or may not be supported by a given switching domain. If it is supported (as indicated through capabilities exchange), every event that involves call control includes a dynamic feature availability parameter that indicates what services are applicable to a given connection, given its current state.

6.9. Status Reporting

Status reporting services are those services used to support the observational portion of the CTI interface's role. If a given computing domain does nothing but observe a switching domain, it uses this subset of services.

6.9.1. System Status

System status services are bidirectional services that are used by each domain to inform the other domain of their current operational status. The *system status* service request contains a parameter that indicates the status of the domain issuing the request. The values that this parameter may take are:

- *Initializing*
- *Enabled*
- *Normal*
- *Message lost*
- *Disabled*
- *Overload imminent*
- *Overload reached*
- *Overload relieved*

If the value is anything other than *normal*, the domain receiving the service request must take appropriate action. System status service requests indicating normal status may be sent periodically by either domain as a *heartbeat* to indicate that it is "alive and well." The positive acknowledgment returned in response by the other domain confirms that it also is "alive."

6.9.2. Snapshot

Snapshot device and *snapshot call* services, as the names imply, allow a snapshot of all information associated with a particular device or call to be obtained.

The snapshot services typically are used when a computing domain is just initializing its view of the switching domain. The *snapshot device* service returns a positive acknowledgment message listing all of the connections associated with a particular device and the state of each. If dynamic feature availability is supported, then the services that can be applied to each connection also is reported.

The *snapshot call* service returns all of the basic call associated information (called device identifier, calling device identifier, associated called device identifier, associated calling device identifier, correlator data) and information regarding each device associated with the call. The per-device information includes the corresponding device identifier, the state of the corresponding connection, and any media services associated with a given connection. If dynamic feature availability is supported, then the services that can be applied to each connection also are reported.

6.9.3. Monitoring

Monitoring refers to the act of requesting that CTI event messages be generated for any changes in state, status, or settings relevant to a particular call or device.

The family of monitoring services includes:

- Monitor Start

    The *monitor start* service requests that monitoring be started for a particular call or device. The parameters for this service include a *monitor filter* that specifies which events, if any, are not desired and should not be sent. This is also referred to as an *event mask* or *message mask* by some.

- Change Monitor Filter

    If the computing domain needs to change the filter (or mask) originally specified, the *change monitor filter* service can be used instead of stopping a monitor and starting a new one.

- Monitor Stop

    The *monitor stop* service is a bidirectional service. If issued by the computing domain, it is a request for the switching domain to clear an existing monitor. If issued by the switching domain, it is an indication that the monitor has been cleared and a request for the computing domain to take any appropriate action.

When the *monitor start* service succeeds in starting a monitor, the positive acknowledgment that it returns includes a monitor cross-reference identifier parameter that uniquely identifies the monitor that was established. All events generated as a result of this monitor are identified with a monitor cross-reference identifier parameter of the same value.

Monitor Objects

Monitor object refers to the item in the switching domain to which a given monitor applies. Monitors can be started on either devices or calls. *Device-object monitors* will exist as long as a given device exists within the switching domain, or until they are stopped by either domain. *Call-object monitors* are stopped when a given call ceases to exist, or if they are stopped earlier.

When a particular call is being monitored (either through call-object monitoring or because it is associated with a device that is being device-object monitored), all events relevant to all connections involving visible devices within the switching domain are reported. Information on connections involving devices not visible within, or outside of, the switching domain may be only partially available or go completely unreported.

Monitor Type

Monitor type determines how calls are monitored when they leave a device. Once a *call-type monitor* has begun reporting events for a particular call for any reason, it continues to report events for that call until the call ceases to exist. In contrast, *device-type monitors* only deliver events relevant to a particular call for as long as they remain associated with a particular device. As soon as it leaves the specified device, monitoring of the call ceases.

The following is the behavior resulting from the four combinations of monitor object and monitor type:

- Device-object, device-type

  This monitor is started on a particular device and reports all events relevant to this device and to calls associated with this device. It continues reporting events relevant to the device for as long as the device exists, but stops reporting events relevant to a given call when that call leaves the device.

- Device-object, call-type

  This monitor is started on a particular device and reports all events relevant to this device and to calls associated with this device. It continues reporting events relevant to the device for as long as the device exists and continues reporting events relevant to a given call for as long as that call exists—even after it leaves the device.

- Call-object, device-type

  This monitor is started on a particular call and continues to report all events relevant to the given call until the call leaves a specified device.

- Call-object, call-type

  This monitor is started on a particular call and continues to report all events relevant to the given call for as long as that call exists.

6.9.4. Device Maintenance

Device maintenance events are a family of events that the switching domain can send to the computing domain to indicate that the functionality associated with a particular device is changing for some reason. These events typically are used to indicate that a device is being manipulated through the OA&M interface or has been physically removed or replaced.

These events include:

- Out of Service

  The *out of service* event indicates that a device has been taken out of service and that events for it may or may not be generated. Service requests cannot be issued for it.

- Device Capabilities Changed

  The *device capabilities changed* event indicates that a device's capabilities have changed and that capabilities exchange service requests should be reissued for the device to determine its new set of capabilities.

- Back In Service

  The *back in service* event indicates that a device is back in service.

6.9.5. Normalized Behavior

Once a monitor has been started for a particular device or call, the switching domain begins sending every applicable event relevant to that entity. A given device or call may be acted upon by service requests from the computing domain or other components in a CTI system, by manual interaction with a telephone set, or by the switching domain itself. The sequence of events that result when a service is invoked for any reason is referred to as the event *flow* for that service. If a standard CTI protocol is being used, the event flow is *normalized*. This means that the behavior seen by any computing domain in the form of an event flow is the same for a given service,[6-15] regardless of how the service was invoked.

If the behavior of a given switching domain is not normalized, the computing domain will be able to track the state, status, or setting of any given telephony resource but will not necessarily be able to interpret any activity that is observed.

6.10. Routing Services

Routing services are among the most powerful of the services provided by a CTI interface. These services permit the computing domain to override the default call routing as determined by the switching domain's call processing resources.

If a computing domain wishes to use routing services, it first uses the capabilities exchange services to determine what devices in a particular switching domain, if any, support routing services. The computing domain then registers[6-16] to use routing services on the device(s) of interest.

6-15. Normalization — Normalization does not apply to vendor specific extensions, which are by definition not standardized in any fashion.

6-16. Routing registration — Certain switching domain implementations do not require registration. They simply assume that the switching domain will want to use routing services. Certain other switching domain implementations allow registering for all the devices in the switching domain at once.

Whenever a call arrives at a device for which routing services are active, a *routing dialog* is initiated by the switching domain. This dialog involves the following steps:

- The switching domain sends the computing domain a *route request* service request, asking it to specify a new destination for a particular call. The *route request* may include a time limit parameter that specifies how much time the computing domain has to make a decision on the routing of the call.

- The computing domain may respond by issuing the *route reject* service to reject the call, the *route end* service to end the dialog (and let the switching domain use its default routing), or *route select* to indicate where the call should be routed.

- If the *route reject* service was specified by the computing domain, the switching domain responds with a *route end* to indicate that the routing dialog has been concluded.

- If the *route select* service was specified by the computing domain and the switching domain found the new destination acceptable, it may optionally issue a *route used* request followed by a *route end* to indicate that the routing dialog is complete. The switching domain might instead respond with a *re-route* service request to indicate that the call could not be routed to the requested destination and the computing domain should repeat the attempt.

Examples of three complete routing dialogs are illustrated in Figure 6-33.

A typical use of routing services is in the distribution of incoming calls to specialists within a team. For example, a travel agency maintains a database of all its clients and each client is assigned to a particular travel agent. When a particular client calls the travel agency, the call is delivered to a hunt group device that ordinarily would just route the call to the next available travel agent. By using routing services, however, the agency's computer is able to override this default

Figure 6-33. Routing dialog examples

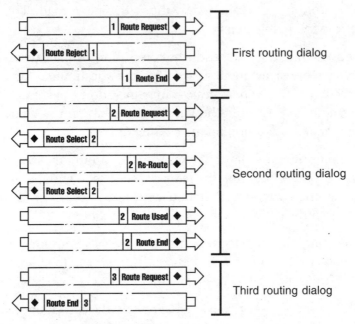

First routing dialog

Second routing dialog

Third routing dialog

routing. It uses the callerID information associated with the call to identify the customer in its database; the call is then routed to that client's assigned travel agent.

6.11. Media Stream Access

Computers have long been capable of dealing with telecommunications data, exchanging faxes, playing and recording sound, and generally working with multimedia data. Excellent technologies, including multimedia architectures and programmatic interfaces for the manipulation of these data types, already exist.

CTI does not supersede, replace, or even encompass these things. Instead, CTI allows existing multimedia technologies to take advantage of a new source of media streams—the telephone network. Switching domain implementations that are capable of providing access to media streams implement support for media services in their CTI interfaces. In fact, media services are the cornerstone of a very large portion of all CTI solutions.

The media-related services provided by a CTI interface are therefore directed at media binding, or accessing the media stream associated with a call in the switching domain, and not media manipulation. Media manipulation is done through corresponding media service interfaces.

6.11.1. Media Stream Access Concepts

A *media stream identifier* allows an association to be established between a given call and the media services available from a particular *media service instance* that is, or can be, associated with the call through a logical device element that supports media access. (Refer to Chapter 3, section 3.7.4 for a discussion of media service instances.)

Media stream access is accomplished through the *attach media service* service request, which is responsible for binding an appropriate media service instance to the call (if available) and returning a media stream identifier. The *detach media service* service request clears the association (and reverses any actions taken by the corresponding *attach media service* service request).

The desired media service instance may or may not be associated with one of the devices already participating in the specified call.[6-17] For example, if the media service instance exists within a modem peripheral attached to a line corresponding to a particular logical device already associated with a given call, then it already has access to the call's media stream. The same is true if the media service instance exists within the switch itself and can be associated with the call transparently. On the other hand, if the media service instance

---

6-17. Multiple media service instances — Depending on the implementation, if there are multiple media service instances available, the computing domain may be able to choose which one is preferred.

exists in a media server,[6-18] an appropriate logical device associated with the server must be added to the call. (See Chapter 7 for more on the actual configuration of CTI systems that utilize media servers.)

If gaining access to the media service instance requires call control activity, the call may be associated with the logical device providing the media services either by joining the device to the call (conferencing), or by diverting the call to the device (transferring) according to the computing domain's request and depending on the switching domain implementation. These operations, if required, are performed by the switching domain and may be implemented using the *transfer call*, *single step transfer call*, *conference call*, or *single step conference call* service. If any call control activities take place as a result of the *attach media service* service request or the *detach media service* service request, appropriate events are provided to indicate what is taking place.

6.11.2. Media Stream Access Model

The mechanisms for utilizing a given media service using a media stream identifier are determined by the definition of the media service itself and generally are industry or de facto standards. The flow of messages between the computing domain and the switching domain implementations relative to a particular media service are referred to as *media service streams*.

Figure 6-34 illustrates the basic model for media stream access. When the CTI client implementation requests the use of a particular media service, it does so using the CTI protocol flowing through the CTI stream. The logical CTI server binds the media service requested to the call specified and returns the appropriate media stream identifier (through the CTI stream). The media stream identifier then can be used by software running on the logical CTI client to access the media service stream appropriately. This includes determining how the

---

6-18. Media servers — Examples of media servers include standalone voice mail systems, voice processing servers, voice response units (VRUs), interactive voice response units (IVRs), fax-back servers, and modem pools.

communication path for the media service stream is to be established. In most cases the communication path used for the CTI stream is related to the one used by the media service stream in some way (e.g., the same LAN, the same multiplexed RS-232 link, etc.).

Figure 6-34. Media stream access model

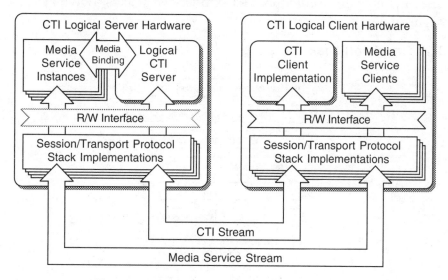

6.11.3. Tone Detection

Tone detection not only is a special form of media access that is intrinsic to the switching domain, it often is used also to detect in-band signaling conventionally associated with the use of media access on voice calls.

The *start telephony tones collection* service requests that the switching domain monitor a given call, or the next call to arrive at a specified device, for telephony tones. If one or more tones are detected, the switching domain generates a *telephony tones detected* event to indicate that the tones were detected. The *stop telephony tones collection* service indicates that an outstanding tone collection request should be canceled.

If a fax calling tone (fax CNG) or modem calling tone (modem CNG) is detected, it indicates that a fax machine or modem on the call would like to begin exchanging modulated data. The computing domain can then use the *media attach service* to request that the appropriate type of modem media service instance be attached to the call. An example showing how this technique is used to receive a fax is shown in Figure 6-35.

Figure 6-35. Receiving a fax

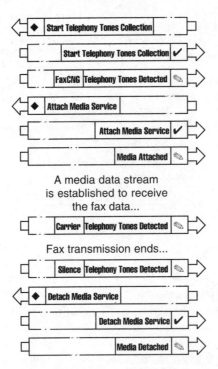

The *start DTMF tones collection* and *stop DTMF tones collection* service requests and the corresponding *DTMF tones detected* event behave in a similar fashion.

6.11.4. Tone Generation

The switching domain is capable of generating telephony and DTMF tones on behalf of the computing domain using the *generate telephony tones* and *generate digits* services respectively.[6-19]

Using these services, a fax transmission can be initiated in much the same way as the fax reception shown in the preceding example.

6.12. Vendor Specific Extensions

Support for vendor specific extensions, often referred to as the *escape mechanism*, is implemented using the *escape* service request and the *private* event. The contents and use of this service and event pair are defined uniquely by every switching domain implementation. These two messages are essentially envelopes into which vendor specific information can be placed for transmission across service boundaries.

To ensure that a computing domain does not use vendor specific extensions inadvertently, the computing domain first must register for them using the *escape register* service (which later can be canceled using the *escape cancel* or *escape abort* services).

Every service and event also carries a *private data* parameter that can be used by a switching domain to extend the functionality of a given service or event in a proprietary way, assuming that this use is negotiated by the computing domain.

The version of a switching domain's vendor specific extensions and private data parameters is independent of the version of the CTI interface being used, so a separate private data negotiation mechanism is used. This involves using the *private data version* service to indicate which of the private data versions supported by the switching domain is to be used.

6-19. Generating tones — It should be noted that whenever tones are generated, the corresponding *tone detected* events are issued by the switching domain even if tone collection is not active. Tone collection applies to tones from other devices on the call.

6.13. Review

In this chapter we have explored the concepts that apply to building and using a CTI interface.

Every component in a CTI system connects to its neighbor(s) through an *inter-component boundary*. As far as each component is concerned, this boundary is a *service boundary* through which it exchanges *CTI messages*. The side of the service boundary that has access to telephony resources is known as the *switching domain* and the side that is seeking to observe and control those resources is known as the *computing domain*. In the relationship between the two components, the one acting as the computing domain is a *logical client* of the one representing the switching domain, which acts as a *logical server*. *Fan-out components* allow the CTI functionality available across one service boundary to be made available to multiple logical clients.

Communication across a service boundary takes place through CTI messages, which may be conveyed through a *CTI protocol* or, if the two components are software running on the same computer, through a *programmatic interface*.

CTI messages contain parameters. *Identifiers* are parameters that make reference to specific telephony resources within the switching domain. Other parameters reflect the actual or desired states, statuses, and settings of resources. *CTI event messages* are sent by the switching domain to provide update information on individual telephony resources. *Service request messages*, *positive acknowledgment messages*, and *negative acknowledgment messages* are used by each domain to request services of the other, and to respond to those requests. An *atomic service* is one that is carried out in one step, and in which success or lack of success is indicated through the type of acknowledgment received. A *multi-step service* is acknowledged positively or negatively when the request itself is validated. In this case, success is indicated only through events that indicate the service completion criteria for the requested service were satisfied. Lack of success is indicated through a special service failure event.

Before any messages can be exchanged between the switching domain and the computing domain, the two first must *negotiate* a CTI protocol (if standard CTI protocols are being used) and the version of the protocol or programmatic interface that is to be used. The computing domain then must use *capabilities exchange services* to determine the telephony resources accessible within a given switching domain, and the capabilities (services and events supported, implementation options, support for media data, etc.) of all devices of interest. The *scope* of a switching domain represents the set of visible telephony resources, that is, the set of telephony resources that may be observed and controlled. Certain resources may not be visible, but their effects within a switching domain can be perceived (e.g., an invisible network interface device or pickup group device). All devices outside the switching domain are part of one or more *external networks*.

In addition to the services for controlling telephony features and services (described in Chapter 4) and services which manipulate all of the components of a physical device element, a switching domain implementation may support *status reporting services*, *routing services*, *media access services*, and *vendor specific extensions*.

The computing domain can observe telephony resources by using the switching domain's *status reporting services*. To the extent that they are available from a given switching domain implementation, these services allow the computing domain to keep track of the operational status of the switching domain (*system status services*), to take a snapshot of a given call or device's status (*snapshot services*), and to request that any change in state, status, or setting associated with a particular device or call be reported through events (*monitoring*). *Device maintenance events* are used to indicate if the capabilities of a monitored device change in some way.

Once monitoring of a particular device has begun, the computing domain receives a *flow* of events corresponding to all of the services that affect that device. If the switching domain implementation supports a standard CTI protocol, the flow of these events is *normalized*. This means that regardless of how a service is invoked, the same event sequence will be observed.

Routing services allow a computing domain to override the default routing of a call from devices that support routing services. *Media access services* allow the media stream of a call to be accessed through a particular *media service instance. Private data*, the *private event*, and the *escape* service request message (often referred to collectively as the *escape* mechanism) allow support for vendor specific extensions that go beyond the functionality provided for in standard CTI protocols and general-purpose programmatic CTI interfaces.

As noted in previous chapters, the concepts presented here do not necessarily reflect the implementation of a given CTI component. They represent a *CTI abstraction* that can be applied to any CTI component in order to integrate that component into a larger CTI system.

7.
CTI System Configurations

This chapter describes how CTI systems are assembled from CTI hardware components, the communication paths that connect them, and the CTI protocols that flow between them. We'll explore the broad range of CTI system configurations that can be assembled with increasingly interoperable CTI system components.

Many different configurations for CTI solutions are presented here, but they are all just examples, or a subset, of the unlimited range of CTI system configurations that are possible. The goal is to provide a collection of building blocks and "serving suggestions" that you can use to plan and implement your own CTI systems.

A CTI system consists of components and communication paths between them. Components may be hardware (switches, CTI servers, telephone stations, PDAs, personal computers, and hybrids) or software (processes running on a hardware component). Messages are passed over communication paths between hardware components (LANs, dial-up, cable, infrared, etc.) or programmatic interfaces between software components.

Graphical Notation for CTI System Configurations

This book uses a standardized graphical notation for describing CTI system configurations. The next two sections describe the general classes of hardware components and the classes of communication paths that can be established between them. Accompanying the description of each component and communication path are the icons and graphics that are used to represent them in CTI system configuration diagrams. These icons and graphics are also summarized on the inside of the back cover.

In general, the icons representing hardware components include any hardware or software necessary for a given hardware component to do its job—to expose a CTI interface and to establish the communication path shown in a particular CTI system configuration. For example, if a component is shown with a modem-based connection to another component, the modems are not shown explicitly. The modem, modem software, serial ports, cables, etc., at each end of the connection are considered part of the corresponding hardware components.

7.1. Hardware Components

In describing CTI system configurations, this chapter deals with discrete hardware components—the individual CTI products—that make up tangible CTI systems. It does not deal with the abstract logical devices that form the basis for the telephony abstraction or the configuration of the software components that are actually installed on the hardware components within a system. (Refer to Chapter 8 for information on the configuration of CTI software within a CTI system.)

This section defines the basic forms that CTI hardware components can take, explains their role in a CTI system, and defines the icons used to represent them. The basic types described can also be hybridized, or combined, to form specialized hardware components. These are described at the end of this section.

7.1.1. Personal Computer

Personal computers are the most pervasive type of computing product today. They are relatively inexpensive computers that range widely in form factor and performance. Several features distinguish personal computers from the other categories of computing products used in CTI systems: They are primarily intended as tools for individual users; they are based on general-purpose, mainstream operating systems; and they support the installation of third-party (add-on) application software and operating system extensions.

For the purposes of a CTI system, a personal computer is considered a *client computer* if it has a software component, referred to as a *CTI client implementation,* that allows application programs to participate as components in the CTI system. (This is described in Chapter 8.)

7.1.2. Multi-user Computer

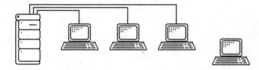

Multi-user computers are like personal computers in most ways except that they are shared by more than one user simultaneously. Each user interacts with the multi-user computer using a terminal or application-specific client-server LAN protocol. The icons above represent multi-user computers and their associated terminals.

Like the personal computer, a multi-user computer which includes CTI client implementation software (see Chapter 8) is considered a client computer for purposes of a CTI system. In general, all client computers are interchangeable in CTI system configurations, so most configuration examples use personal computers.

7.1.3. PDA

A *personal digital assistant (PDA)* is another form of computing product. While it is like a personal computer in the sense that it is intended as a personal tool for an individual, it is quite different in most every other way. PDAs are special-purpose products that typically have only enough computing power and storage capacity to carry out the specific tasks for which they were designed. As a result, PDAs generally have special-purpose operating systems or no operating system (in the traditional sense) at all. PDAs may or may not allow the installation of third-party software.

PDAs are usually in the form of highly portable products such as electronic organizers, electronic clipboards, intelligent wristwatches, and smart remote control units. However PDAs can also be embedded in automobile dashboards, refrigerator doors, and hospital beds.

7.1.4. Telephone Stations

Telephone stations are the devices people use to directly access media streams from telephone networks. The range and variety of telephone stations is even greater than that of personal computers. They include home phones, desk phones, attendant consoles, cellular phones, pay phones and even fax phones and video phones. Refer to Chapter 5 for more information about telephone station products.

If the telephone station provides a CTI interface, the switching domain that can be accessed (depending on the implementation) may encompass the entire switch to which it is attached, or it may treat the switch as an external network and provide control over just the lines (logical devices) that are directly connected. The CTI interface may or may not provide access to the associated physical device element.

Typically a telephone station will expose a CTI interface in one of two ways:

- Built-in CTI interface
  Some telephone stations, so-called *smart phones*, have the necessary built-in intelligence and communications hardware to allow a CTI client to connect directly to the physical telephone set itself.

- Add-on CTI interface
  Some telephone stations do not have a built-in CTI interface but were designed with a CTI interface as an option. If available, these modules allow a separate CTI interface component to be added via a snap-on module, upgrade card, or some other mechanism. This add-on exposes an interface for use with a standard communication path (as described in section 7.2).

These CTI interface variations often provide media services of some type, most notably asynchronous data and possibly isochronous digital data or audio. They also may have built-in data or fax modem functionality.

7.1.5. Telephone Station Peripheral

A *telephone station peripheral* is a computer peripheral that can be interfaced to one or more telephone lines in place of, or in addition to, a telephone station.

If the type of line involved is analog, ISDN, or another nonproprietary type, the telephone station peripheral may be from any vendor and the product is likely to be in the form of a common modem or ISDN terminal adapter. If the type of line is proprietary, however, the switch vendor generally is the exclusive source of any telephone station peripheral that can be connected. These peripherals may be attached to a computer system through an external connector, or may be in the form of an add-in card.

Telephone station peripherals almost always provide some level of media access service. Typically these peripheral products either include or provide access to data and fax modem functionality, and may also provide access to audio streams and raw isochronous data.

If the telephone station peripheral provides a CTI interface, then (depending on the implementation) the switching domain that can be accessed may encompass the entire switch (or network) to which it was attached, or it may treat the switch as an external network and provide control over just the lines (logical devices) that are directly connected. The peripheral may or may not be able to control and observe the physical device element(s) associated with telephone station(s) on the same line(s).

7.1.6. CTI Server

In a CTI system configuration, a *CTI server* is a computer of some sort that is dedicated (from the perspective of the CTI system) to playing the role of a logical CTI server to a set of logical clients in the form of client computers and PDAs. The CTI server hardware may be any type of computing platform running any type of operating system. The only essential characteristics are that it physically has the ability to connect to its logical server and logical clients and that it supports a software component known as a CTI server implementation which carries out its role as a CTI server. (CTI server implementations are described in section 7.3 and in Chapter 8, section 8.2.)

A CTI server generally plays one or more of the following roles in a CTI system:

- A CTI server is usually a fan-out component that channels CTI control and status messages between its logical clients and logical server.

- A CTI server also may act as a secure gatekeeper for its logical server's CTI interface. A switch that does not authenticate the clients that connect to its CTI interface can be front-ended by a CTI server that provides this capability. The security added may be limited to restricting access to authorized clients, or it may include customizing the view of the switching domain that each client sees.

7.1.7. Media Server

A *media server* is a computer of some sort that is dedicated (from the perspective of the CTI system) to providing media services to a set of logical clients in the form of client computers, PDAs, and CTI servers. A media server works in conjunction with an associated logical CTI server. It is invoked when the logical clients of the CTI server use the appropriate media access services (described in Chapter 6) for requesting access to a particular media service stream.

Media servers are considered part of the switching domain by logical clients but are actually considered telephony resource to whichever component in the chain of CTI components is providing the media access binding function.

7.1.8. CO Switch

For purposes of this chapter, a *CO (central office) switch* is any switch operated by a common carrier (i.e., telephony company). It is typically located in one of the carrier's central offices and may provide wireline service, wireless service, or both. (See Chapter 5 for more information.)

7.1.9. CPE Switch

A *CPE (customer premise equipment) switch* may be a front-end switch, a KSU,[7-1] a PBX, or an application-specific switch; it may support telephone stations through wired (analog, ISDN, proprietary) or through wireless connections. (Refer to Chapter 5 for more information on the forms of CPE switch products.)

7.1.10. Internet Voice Gateways

Internet voice gateways are a special type of switch that has network interface devices for both switched telephony and Internet telephony. They can route calls between points on the Internet's virtual voice network and the worldwide telephone network.

Internet voice gateways may be substituted for switches in any of the configurations described in this chapter.

7-1. Hybrid switch — In some cases a CPE switch may be a so-called *hybrid switch* that is a combination PBX/KSU, but these are generally treated as PBXs.

7.1.11. Hybrids

Hybrid components combine the features of two or more basic hardware components. Described below are a few of the most compelling hybrids that may be formed by blending hardware components described previously. Not all the hybrids described here are explicitly featured in the CTI system configurations in this book, however they can be constructed by simply making appropriate substitutions.

- Client computer + CTI server

    The hybrid product equivalent to a client computer and a CTI server is a single computer that is able to both run CTI applications and provide fan-out of CTI services to other logical clients in a CTI system. The dual role of this computer is transparent to CTI software components running on it so from their perspective, this hybrid is equivalent to a client computer. Configurations featuring this type of hybrid can be constructed by substituting it for a CTI server. (An example using this type of hybrid is presented in Chapter 9.)

- Media server + CTI server

    The hybrid product equivalent to a media server and CTI server is a single hardware component in which both types of servers are simultaneously active. A component of this type is effectively a CTI server that is able to provide media access services to its clients more easily by incorporating the media services rather than by binding to them in another, independent server. (Examples of this type of hybrid are discussed in section 7.6.3 and in Chapter 9.)

- Switch + CTI server

    A hybrid product with the functionality of a switch and a CTI server is a switch of some sort that provides the connectivity associated with a CTI server and also includes a CTI server implementation software component. This is a compelling hybrid because it provides the scalability associated with a CTI server, while simplifying the configuration by substituting for a switch-CTI server combination in any given system.

- Switch + media server

  The hybrid product with the functionality of a switch and a media server is a switch of some sort that provides media access services to its clients more easily by incorporating the media service resources rather than by binding to them using an external media server. This type of hybrid is compelling because it eliminates the complexity associated with connecting an independent media server to a switch. CTI configuration using hybrids of this type can be constructed by substituting for any switch-media server combination.

- Internet telephony gateway + media server

  The hybrid product that is equivalent to an Internet telephony gateway and media server is a computer that is very similar to the switch-media server combination described above. The same physical hardware is required to build a media server and an Internet gateway; the only difference is in the software that they run. This hybrid is therefore a natural and likely combination because it involves simply running both types of software on the same machine. A hybrid of this sort can be substituted for any Internet telephony gateway (which in turn can be substituted for any switch).

- PDA + telephone station

  Hybrid products with the functionality of a PDA and a telephone station are sometimes referred to as "intelligent telephones" or even "intelligent communicators" and can be found in both desk set and wireless formats. They effectively embed some level of personalized computing functionality in a hardware component that is otherwise a telephone station. From the perspective of CTI system configurations, they are functionally equivalent to a telephone station and can be substituted accordingly. From the perspective of any software components running inside the PDA portion of the product, there is effectively a PDA inside the telephone station in a direct-connect relationship with the telephone station portion.

7.2. CTI Communication Paths

CTI communication paths, or *CTI links*, are the data connections that two hardware components use to establish an inter-component boundary with one another in a CTI system.

A CTI communication path consists of both the physical connectivity (cabling, hardware transceivers, etc.) and the session/transport protocol stack[7-2] used above it.

7.2.1. CTI Stream and Media Service Stream

The relationship between the two hardware components on each side of a particular communication path is illustrated in Figure 7-1.

Figure 7-1. Anatomy of a communication path

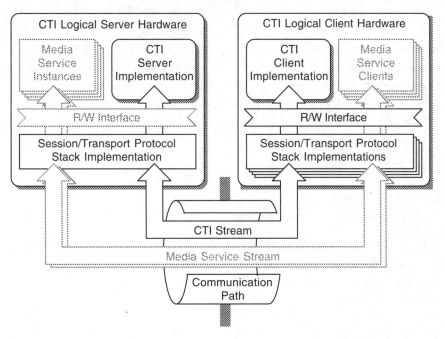

7-2. Session/transport protocol stack — In ISO terminology the *session/transport protocol stack* consists of data link, network, transport, and (optionally) session layers (2 to 5). A *communication path* represents ISO layers 1 to 5.

401

The communication path between two hardware components is used to carry CTI messages (service requests, events, etc.) between the logical CTI client and the logical CTI server that exist on each side of the service boundary that the communication path represents. This stream of messages is referred to as the *CTI stream*.

If the switching domain implementation in the pair of components includes one or more media services, and the logical CTI client uses media access services to bind to one or more of them, the resulting *media service stream(s)* may also be carried by the same communication path, if appropriate.

Note that in Figure 7-1 the computing domain implementation is shown as having multiple session/transport stack implementations, while the switching domain is shown as having only one. This generally is the case, but there may be some exceptions. Most switching domain implementations have proprietary internal structures, so the notion of a R/W interface[7-3] really is not applicable. By the same token, they tend to implement support for just one type of communication path (preferably one supported by all computing domain implementations). On the other hand, computing domain implementations typically are based on operating systems that provide R/W interfaces and multiple protocol stacks.

7.2.2. LAN

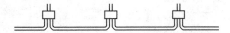

A *LAN* (*local area network*) communication path uses a LAN protocol stack and multi-point communication infrastructure to establish streams between CTI components.

7-3. R/W interface — R/W interfaces are programmatic interfaces that provide access to session/transport protocol stack implementations. They are described in Chapter 8, section 8.3.1.

The physical network used for LAN communication paths in a given configuration may be wired or wireless, may have any topology (structure), and may support thousands of components or as few as two. Components may be directly connected to the LAN or may be indirectly connected using a dial-up bridge. The key feature characterizing a LAN's physical layer is that it represents a shared medium over which any number of connected components can setup, tear down, and simultaneously use, communication paths with any number of other components.

LAN communication paths may be established using any session/transport protocol stack that provides reliability and virtual circuits (sessions) of some sort. The stack optionally may support authentication and encryption. Examples of the popular LAN protocols satisfying these requirements include:

- TCP/IP (Internet protocol family)

- IPX/SPX (Novell's protocol family)

- ADSP (Apple's AppleTalk protocol family)

The graphical notation for a LAN communication path is shown above (symbolically it appears as a daisy chain although it might be any topology). Figure 7-2 shows a LAN communication path between a personal computer and a CTI server.

Figure 7-2. LAN connection

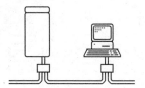

7.2.3. Serial Cable and Serial Bus

Serial cable and *serial bus* communication paths are used to establish streams between two CTI components in close proximity. At the physical level, these are inexpensive cables that carry serial data over short distances. One serial cable may constitute the whole physical layer or a single segment in a serial bus.

Serial options include:

- RS-232 (or V.24)
  The universally supported point-to-point serial interface.[7-4]

- USB (Universal Serial Bus)
  Multi-point/multi-channel communication

- GeoPort
  Multi-channel isochronous communication

- FireWire (IEEE 1394)
  Multi-point/multi-channel isochronous communication

The graphical notation for a serial cable/bus communication path is shown above (symbolically it appears as a single span although it might be part of a multi-point serial bus). Figure 7-3 shows a serial cable/bus communication path between a multi-user computer and a media server.

Figure 7-3. Serial connection

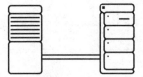

7-4. Multiplexed serial — The *Versit CTI Encyclopedia* specifies a number of mechanisms that can be used to reliably multiplex CTI and media service streams over an RS-232 or equivalent connection in a standard fashion. This includes a specification for implementing standard CTI protocols in products that also operate using the Hayes AT command set.

7.2.4. Infrared (IR)

Infrared (*IR*) communication paths are used to establish streams between two CTI components without requiring cables. At the physical level, the connection is established by infrared light transmitters and receivers that are built into each component. Infrared signaling, long used for remote controls on consumer electronics products, is now a proven technology for data exchange, particularly between mobile products such as notebook computers and PDAs.

The Infrared Data Association (IrDA) has defined standard IR protocol stacks for interoperability between the products of different vendors.

The graphical notation for an infrared communication path is shown above and Figure 7-4 shows its use in a configuration involving a PDA and a pay phone.

Figure 7-4. Infrared connection

7.2.5. Dial-up

A *dial-up* communication path is one that takes advantage of a wide area telecommunications network (*WAN*). (See the sidebar "Telephony and Telecommunications" on page 68 for more information.) As with the other types of communications paths, once a dial-up communication path has been established between two points using one of these well-defined standards, the communication path operates as if the two components were directly connected using a serial cable.

Typical dial-up communication paths include:

- Modem connections (modulated data)

- SVD modem connections (supporting simultaneous voice and data)[7-5]

- Digital communication paths (such as ISDN)

- Packet data virtual circuits (such as X.25)

The particular telecommunications technology being used determines how many dial-up communication paths a single physical connection to the network (e.g., telephone line and modem, ISDN line and TA, X.25 PAD, etc.) can support.

The graphical notation for a dial-up communication path is shown above. Figure 7-5 shows its use in a configuration involving a notebook computer and a CTI server.

Figure 7-5. Dial-up connection

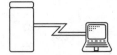

7.3. CTI Streams and CTI Protocols

CTI protocols define the structure, contents, use and flow of PDUs carrying CTI messages for a given CTI stream. CTI protocols are high-level protocols[7-6] like the protocols used to send electronic mail, print to a printer, retrieve files from a file server, or browse the World Wide Web. Like these other protocols, they are designed for transmission

7-5. Simultaneous voice and data — Simultaneous voice and data (SVD) involves using modulation and compression technologies to allow both voice and other media data to travel simultaneously over a single voice media stream channel.

7-6. ISO layering — In ISO terminology, CTI protocols are application layer protocols and their definitions include the presentation layer encoding. CTI protocol definitions correspond to ISO layers 6 and 7.

over a reliable, connection-oriented (i.e., guaranteed-delivery) session/transport protocol stack referred to as a *communication path*. (Refer to the previous section for more detail on communication paths.) CTI protocols may travel between any two CTI components (hardware or software) across any appropriate communication paths. While CTI protocols represent the only way to convey CTI messages between two hardware components they are equally applicable to two software components within a single hardware component.

A CTI component capable of interoperating with other components using CTI protocols must include a module known as a *CTI protocol encoder/decoder*. As shown in Figure 7-6, a CTI protocol encoder/ decoder is responsible for managing CTI streams (setting up communication paths, sending and receiving PDUs, etc.) and for encoding and decoding the PDUs based on knowledge of the CTI protocol in use. When implemented as a software module running on an open computer system, CTI protocol encoder/decoders utilize a system's R/W interface in order to access the session/transport protocol stack used to deliver the CTI stream. (Refer to Chapter 8 for more on software configurations.)

Figure 7-6. CTI component using CTI protocols

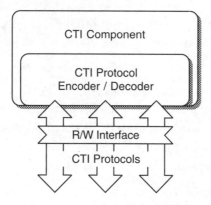

A given interoperable CTI component may be either a logical server, a logical client, or both a logical server and a logical client simultaneously. (These concepts are illustrated in Chapter 6, section 6.3.) A hardware component that acts as a logical client with respect to one component and as a logical server to one or more logical clients contains a software component known as a *CTI server implementation*. As illustrated in Figure 7-7, a CTI server implementation interprets and keeps track of all the CTI service request, event, and acknowledgment messages from both sides and, as appropriate, conveys them to their correct recipient.

Figure 7-7. CTI server implementation component

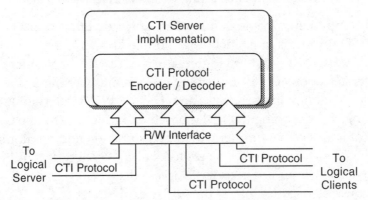

For example, if a particular logical client sent a CTI service request message, the CTI client implementation would pass the service request to the logical server, receive the acknowledgment, and forward the acknowledgment to the logical client that originally made the request. In this fashion a CTI server implementation acts as a proxy[7-7] for its logical clients.

If the CTI protocol encoder/decoder in a given CTI server implementation supports multiple CTI protocols, it is able to effectively act as a translator between logical clients using one protocol and a logical server using another. (Refer to Chapter 8, section 8.2. for more information on the features and responsibilities of CTI server implementations.)

7-7. Proxy server — The operation of a CTI server implementation can be compared to the operation of a proxy-based Internet firewall implementation. Both are servers that appear to be providing a certain service to their clients; in reality they access the actual service on behalf of those clients.

7.3.1. Standard CTI Protocols

CTI Plug & Play interoperability is made possible through the use of standard CTI protocols. CTI Plug & Play means that no software specific to a particular logical server needs to be installed on a logical client. This type of interoperability between any two CTI components is achieved when each CTI component implements both:

- Standardized CTI protocol(s)

- Standard communication path(s)

A CTI component that does not satisfy both of these requirements requires special software to be installed and thus is not CTI Plug & Play. (For more information on proprietary implementations and software configurations, see the next section and Chapter 8 respectively.)

Three standard CTI protocols have been defined in the Versit CTI Encyclopedia.[7-8] They are:

① "CTI Protocol 1"
CTI Protocol 1 is optimized for use between a switch and a CTI server

② "CTI Protocol 2"
CTI Protocol 2 is optimized for use between client computers and CTI servers

③ "CTI Protocol 3"
CTI Protocol 3 is optimized for use between telephone stations, PDAs, and client computers

The Versit-defined protocols all represent the same application layer protocol, but they are optimized for different implementations through the use of different encodings, or presentation layer protocols.[7-9] Each CTI protocol may be used across any service boundary, and each is capable of carrying the same information,

---

7-8. Versit CTI Encyclopedia — *Versit Computer Telephony Integration (CTI) Encyclopedia* (Versit, 1996).

though they have been optimized for use with specific types of CTI system components. For example, CTI Protocol 3 is optimized to use buffers as small as 80 bytes, which is necessary for implementing CTI components such as CTI-enabled pay phones, cellular phones, desk phones, and PDAs. In contrast, CTI Protocol 1, which is intended for a dedicated link between the CTI interface on a switch and a CTI server, uses ISO's ASN.1 encoding and requires buffers that may be as large as 2000 bytes.

Graphical Notation for Standard CTI Protocols

Communication paths indicated in the CTI system configuration diagrams in this book carry CTI streams using standard CTI protocols unless indicated otherwise (as described in section 7.3.2). The flow of a standard CTI protocol between various components using different communication paths is indicated as shown in Figure 7-8.

Figure 7-8. Standard CTI protocols traveling over communication paths

When indicating the use of a specific standard protocol for a particular communication path in a CTI system configuration, the protocol's number is used as illustrated in Figure 7-9. In this example a LAN communication path is shown carrying a CTI stream using CTI Protocol 2.

Figure 7-9. Example CTI protocol

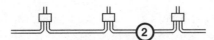

7-9. Application and presentation layer protocols — In the ISO model, the presentation layer (ISO layer 6) is concerned with the syntax and semantics used to encode application layer data. The application layer (ISO layer 7) is concerned with conveying parametrized information reflecting the abstraction of some service or capability.

7.3.2. Proprietary CTI Protocols

The presence of a proprietary CTI protocol traveling over a communication path in a CTI system configuration diagram, is indicated by shading the appropriate communication path symbol in the fashion illustrated in Figure 7-10.

Figure 7-10. Proprietary CTI protocols traveling over communication paths

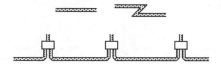

As described in Chapter 1, section 1.6, CTI products based on proprietary CTI protocols take two forms, corresponding to the first two phases of CTI evolution:

- Custom CTI systems

- API-centric products

In addition, CTI products using proprietary CTI protocols may be made to interoperate with CTI components that are based on standard protocols by using *mappers*.

Custom CTI Systems

CTI components relying on proprietary protocols may be formed into a closed, custom CTI system that can be viewed only as a single, turnkey CTI solution. All the CTI components making up the solution must be from the same vendor or be developed to that vendor's specifications. This is illustrated in Figure 7-11.

All the software and hardware making up such a CTI system is interdependent, so the solution can be used in only the configuration or configurations for which it was developed. The vendor of such a system may base its configuration on any of the system configurations described in this chapter, but system integrators and customers are not in a position to reconfigure, expand, or add to the system.

Figure 7-11. Custom CTI solution using proprietary CTI protocol

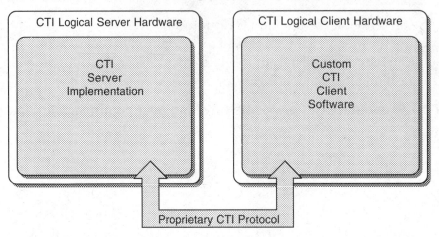

API-centric CTI Products

Another approach for building CTI solutions involving proprietary CTI protocols is through the use of special pieces of *API-centric adapter software*[7-10] that are able to establish appropriate communication paths, interact with the proprietary protocol, and make their functionality accessible through a programmatic interface (typically a procedural API). These special pieces of code must be written by or for the vendor of the proprietary protocol, and must be developed and tested for each operating system platform and system model to which the CTI component is to be connected. CTI hardware components for which the vendor chooses not to develop this special software, and those that do not have standard APIs (or operating systems), have no access at all to CTI components that use proprietary protocols. Depending on the platform, this special piece of adapter software is referred to as a *driver*, *service provider*, or *telephone tool*. This is illustrated in Figure 7-12.

7-10. API-centric adapter software — An implementation of API-centric adapter software is a type of CTI client implementation in the CTI software framework that is presented in Chapter 8.

Figure 7-12. Proprietary CTI protocol with adapter software

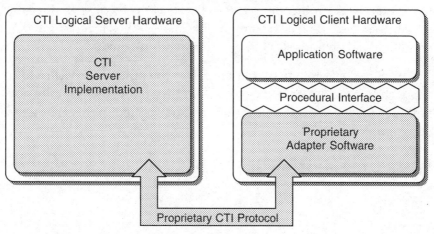

The CTI component with the proprietary CTI protocol and its associated adapter software are interdependent, so the CTI component in question can be used only in the configuration or configurations for which it was developed. The vendor of such a system may base its configuration on any of the system configurations described in this chapter, but system integrators and customers are not in a position to reconfigure, add new platforms, or upgrade these CTI systems without first obtaining the appropriate adapter software (if available).

The significant improvement that this represents over custom CTI systems is that off-the-shelf applications that are compatible with the particular API chosen (and the behavior of the proprietary protocol under that API) may be used instead of custom-developed application software.

Protocol Mappers

CTI component implementations designed for CTI Plug & Play interoperability generally will support only standard CTI protocols across the R/W interface (refer to Figure 7-6). However, a CTI component that is not CTI Plug & Play because it implements only proprietary protocols still can achieve interoperability by providing a *protocol mapper*, or *mapper*, for short.

Mappers are software or hardware components that transform a proprietary CTI protocol into a standard CTI protocol so that it can be used by a standards-based component. They do this in a fashion that is entirely invisible to the CTI Plug & Play component.

The principal advantage of mappers over the other approaches to supporting proprietary protocols is that they allow for virtually any arbitrary CTI system configuration (as exemplified in this chapter) to be assembled.

Protocol Mapper Hardware

Hardware mappers are physical adapters that sit between the two components in question and are transparent to both. The icon for a hardware mapper is shown above and an example of a CTI configuration involving a protocol mapper is illustrated in Figure 7-13.

Figure 7-13. Hardware mapper example

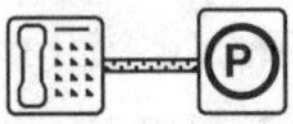

Despite the incremental cost associated with an extra piece of hardware in a given configuration, hardware mappers generally are a more attractive option than software mappers because only one version needs to be developed. This single implementation will work with any CTI Plug & Play component—in contrast to the software mappers, described below, which must be developed for each and every CTI component platform. (For some types of components, such as an IR-based remote control, this is impossible.) It is also an attractive approach for many vendors because it allows them to incorporate the hardware mapper into the CTI product in its next generation.

Protocol Mapper Code

CTI components that are not CTI Plug & Play and do not support hardware mappers need to provide *mapper code*, or software, that runs on each of the CTI components with which they need to interoperate.[7-11] Software mappers are invisible to CTI software components because they are implemented as session/transport protocol stack modules. So, like hardware mappers, they appear to be within the communication path itself. (Software mappers are discussed further in Chapter 8.)

Although conceptually a piece of protocol mapper software is very similar to the adapter software used in API-centric models, it has two key advantages. The first is the ability to flow the resulting standard protocol to any number of downstream CTI Plug & Play components using a fan-out component. This means that a CTI product with proprietary protocols can be used in a much broader range of configurations. An equally significant advantage of mappers over API-centric adapter software is that mapping to a standard protocol causes CTI behavior to become normalized, which in turn eliminates restrictions on the range of applications that work reliably using that particular API-CTI component combination. One disadvantage software mappers share with API-centric adapter software is the fact that neither can support logical CTI clients that lack traditional operating systems (such as PDAs and consumer electronics products).

In the CTI system configuration diagrams in this book, the presence of mapper code installed on a given CTI component to map from a proprietary CTI protocol is indicated with the symbol shown above.

7-11. Mapper economics — The economics of developing mapper code might not seem evident at first glance, as it requires that mapper code be written and tested for many different platforms, operating system versions, and hardware configurations. Some vendors have determined it to be a faster migration route, however.

7.4. Media Service Streams

Media service streams carry media service data over a communication path between a component containing a media service instance to a corresponding media service client (Figure 7-1). In the graphical notation for CTI system configuration diagrams, the presence of a media service stream on a particular communication path (PDUs encoded according to the corresponding media service protocol) is indicated with the symbol shown above.

Media Service Instances

Media service instances may be implemented on any type of hardware component described in section 7.1, not just on dedicated media servers. The presence of a media service instance in a given CTI component is indicated with the symbol above. All logical clients of the logical server supporting the media service instance see a switching domain that includes this media service. In Figure 7-14 a personal computer accesses the media service instance in a telephone station.

Figure 7-14. Media service stream traveling over a communication path

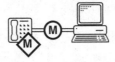

Media Service Mappers

If the media service instance is located in a component other than the associated CTI component, or if the CTI stream and media service streams are not delivered over an associated communication path, media access binding must be implemented using *media service mapper code*. The presence of a media service mapper is indicated with the symbol shown above.

By simultaneously establishing communication paths both to a media server using an appropriate media service stream and to a logical CTI server using standard CTI protocols, this type of mapper creates the illusion of a switching domain that includes media access capability. As a mapper, it tracks the messages traveling in either direction on both streams and presents them to its own logical clients in an integrated fashion.

7.5. Direct-Connect Configurations

Direct-connect configurations, as the name implies, involve a direct connection between a user's client computer or PDA and a telephone station or telephone station peripheral.

All of the configurations in this section are presented using the standard graphical notation. Refer to the inside of the back cover for a summary of the symbols.

7.5.1. Basic Direct-Connect Configuration

In a *direct-connect configuration*, a client computer or PDA is connected directly to a telephone station or telephone station peripheral that supports a CTI protocol for access to CTI services. The logical client is connected using a serial cable (RS-232), serial bus (FireWire, GeoPort, or USB), infrared communication path, or a product-specific communication path such as an add-in card for a particular computer architecture. Figure 7-15 shows an example of a CTI Plug & Play direct-connect configuration.

Figure 7-15. Direct-connect configuration example

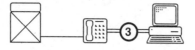

Telephony features and services available through the resulting CTI stream may be restricted to just the logical and/or physical device elements corresponding to the connected telephone station. This is referred to as *first-party call control* and involves limiting the scope of the switching domain to just a single device or device configuration. This is illustrated in Figure 7-16. If the switching domain contains additional telephony resources, it is referred to as *third-party call control*. The switching domain in this case may consist of station devices and/or additional telephony resources within the switch in addition to the connected telephone station, as shown in Figure 7-17. In either case, the CTI stream from the switch is delivered through a communication path with a telephone station or telephone station peripheral.

Figure 7-16. Direct-connect first-party call control

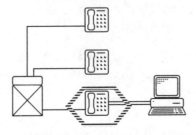

Figure 7-17. Direct-connect third-party call control

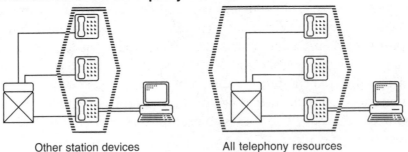

Other station devices   All telephony resources

7.5.2. Direct-Connect Mapper Configurations

If the telephone station or telephone station peripheral does not implement a standard CTI protocol, it may have appropriate mapper hardware or mapper code available; otherwise, the telephone station must be used in conjunction with API-specific adapter code or custom application software.

Direct-Connect Mapper Hardware

Figure 7-18 presents an example of a CTI system configuration involving a piece of direct-connect *mapper hardware*. This special hardware component (labeled "P") sits between a telephone station that employs a proprietary CTI protocol and a client computer that supports standard CTI protocols. In this particular example, the mapper hardware supports CTI Protocol 3 and the telephone station's proprietary protocol and translates between the two.

Figure 7-18. Protocol mapper hardware configuration

As shown here, a direct-connect hardware mapper is a miniature CTI server of sorts that can be installed between two otherwise incompatible CTI components. The combination of the telephone station and the protocol mapper hardware is functionally equivalent to a single CTI Plug & Play telephone station (except to those in the CTI value chain that must obtain, install, and manage two components instead of one). In this example the hardware mapper is connected using a serial cable or bus to connect to each side, however both communication paths need not be the same. For example, the mapper hardware could be designed to connect to the telephone station using a serial link and to the client computer or PDA using infrared.

Direct-Connect Mapper Code

Figure 7-19 presents examples of proprietary *mapper code* being used in direct-connect configurations. The telephone stations in these configurations implement a proprietary CTI protocol. In order to establish a CTI stream with them, each of the different client computers and PDAs must use a piece of mapper code (labeled "P") that corresponds to the proprietary protocol for the telephone stations being connected.

Figure 7-19. Direct-connect protocol mapper code configuration

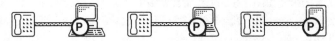

The disadvantage associated with mapper code is that it must be developed for every popular version of every popular operating system (e.g., DOS, Mac OS, Newton OS, Pen Windows, Unix, Windows, Windows NT, etc.). It also assumes that a given client product will actually have an operating system and a CTI client implementation[7-12] that supports the installation of mappers.

7.5.3. Direct-Connect Media Access Configurations

Media access of some sort is quite common in direct-connect configurations because the communication path used for delivering the CTI stream often can be shared easily for delivering a media service stream as well.

CTI Plug & Play Media Access

Figure 7-20 shows a CTI configuration in which a telephone station provides access to a media service instance through its CTI interface. The logical client in this example, a personal computer, is using

7-12. CTI client implementation — CTI client implementations are CTI software components in the CTI software framework that is presented in Chapter 8.

Figure 7-20. Direct-connect CTI Plug & Play media access configuration

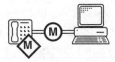

standard CTI protocols for call control, etc. After requesting media service binding a media service stream was formed over the same serial communication path used by the CTI stream.

Mapper Hardware Media Access

In the system configurations presented in Figures 7-21 and 7-22 media access functionality has been added to the basic mapper hardware configuration previously presented in Figure 7-18.

Figure 7-21. Mapper hardware media access configuration

Figure 7-22. Mapper and media access hardware configuration

In the example shown in Figure 7-21, the media service instance is in the telephone station and the telephone station's proprietary protocol mapper supports media access services using proprietary messages which are translated by the mapper. In contrast, Figure 7-22 shows a case where the media service instance is actually in the mapper hardware itself. In the latter case, mapper hardware has access to the isochronous data streams associated with the telephone station and is therefore able to provide the media services itself.

Mapper Code Media Access

Figures 7-23 and 7-24 correspond to the mapper software versions of the mapper hardware examples shown in Figures 7-21 and 7-22 respectively. In Figure 7-23 the telephone station provides the media service instance so the mapper code simply translates protocols appropriately.

Figure 7-23. Mapper code media access configuration

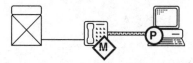

Figure 7-24. Mapper and media access code configuration

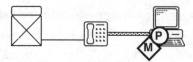

In the second case, the media service instance is actually implemented within the mapper code running on the client computer. The communication path to the client computer is FireWire, GeoPort, or an equivalent technology that permits direct access to the isochronous data streams associated with the telephone station. The mapper code is then able to terminate the media stream itself, in addition to providing CTI protocol mapping.

7.5.4. Smart Phone Serial Cable/Bus Configurations

A *smart phone* is a telephone station that has built-in or add-on CTI interfacing (i.e., it is "smart"). A client computer or PDA connects to a smart phone using a serial cable (RS-232), GeoPort, FireWire, or USB serial bus, as shown in Figure 7-25. In this configuration, standard CTI protocols are supported by the telephone stations so no proprietary mapper code is present.

Figure 7-25. Smart phone serial cable/bus configuration

This is a likely configuration for office and home situations. It also is a likely configuration for instances where a notebook computer is docked for the night with a hotel phone or is connected to a cellular phone or the seat-side phone in an airplane or train.

Mappers may be used as described in section 7.5.3, but are practical only in home or office situations. Mappers are not practical for mobile users who are likely to encounter many different telephone products from many different vendors during the course of each day and who cannot acquire and install special software for even a fraction of them.

7.5.5. Smart Phone Infrared Configuration

Infrared (IR) is a significant and important alternative to serial cables for connecting CTI products. It allows mobile products to connect and disconnect without requiring physical contact. This is critical for CTI solutions in any kind of public setting, such as an airport, hotel lobby, or vehicle. IR also is a very attractive communication path in home environments, where CTI-enabled remote control units can interact with the CTI-enabled functionality of telephone stations built into consumer electronics products such as televisions. Finally, IR used in office environments allows for instant docking of notebook computers and PDAs with telephone stations in conference rooms, lobbies, and work areas.

Two examples of IR-based CTI system configurations are shown in Figure 7-26. Here an IR communication path is used to connect a PDA with a pay phone and a laptop computer is connected to a desk set telephone station.

Figure 7-26. Smart phone infrared configuration

Telephone stations supporting IR tend to be in high traffic areas and in contexts where people will establish and tear down communication paths with these products frequently. In most cases, mappers which must be specially installed and configured are not practical for IR-based products.

7.5.6. Serial-Based Telephone Station Peripheral Configuration

In this direct-connect, serial-based CTI configuration, a telephone station peripheral providing a CTI interface connects to one or more lines from a switch. There may or may not be one or more telephone stations on the same line and, depending on the implementation, the CTI interface may or may not be able to observe or control their physical elements. As shown in Figure 7-27, the peripheral may be attached by a serial cable or serial bus using a standard raw serial stream. The configuration is CTI Plug & Play and requires no special software on the client computer or PDA.

Figure 7-27. Serial telephone station peripheral configuration

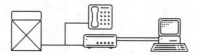

Inexpensive, so-called *dumb peripherals* deliver raw data streams from the telephone line across the serial interface. They require installation of appropriate mapper code on the client computer or PDA to translate these raw data streams into independent CTI protocol streams and media service streams. This configuration is shown in Figure 7-28.

Figure 7-28. Mapper code serial telephone station peripheral configuration

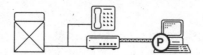

Another variation involves having a telephone station in tandem, so that the telephone station peripheral is between the switch and the telephone station (Figure 7-29). This arrangement gives the peripheral greater control over the telephone station. Depending on the implementation, the peripheral can take over the role of the switch controlling and interpreting commands from the telephone station.

Figure 7-29. Tandem serial telephone station peripheral configuration

7.5.7. Add-In Board Configuration

The *add-in board telephone station peripheral* configuration is very similar to the serial-based telephone station peripheral configuration. The only difference is that, rather than being connected to the client computer through a serial cable or serial bus, the peripheral is in the form of an EISA, ISA, VESA, MCA, NuBus, PCI, or PCMCIA add-in card.

Because these boards generally do not have an associated session/transport protocol stack, they usually require a mapper if standard CTI protocols are to be used. In most cases, however, the mapper is a simple virtual serial port implementation that allows the CTI interface on the board to be accessed as if it were the type of serial-based peripheral described in section 7.5.6. This mapper-based configuration is shown in Figure 7-30.

Figure 7-30. Mapper code add-in board configuration

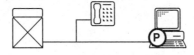

One case where a mapper is not required is the case of a PCMCIA implementation that uses the same mechanism as a modem card to expose a serial interface. In this case the configuration is CTI Plug & Play, as shown in Figure 7-31.

Figure 7-31. CTI Plug & Play add-in board configuration

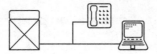

As with the serial-based peripheral, the add-in board also may be arranged in tandem with a telephone station. This is most notably the case where the add-in card attaches to a digital line and provides an analog line interface. This is illustrated in Figure 7-32.

Figure 7-32. Tandem mapper code add-in board configuration

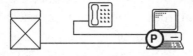

Add-in board configurations are applicable primarily in scenarios where the line interface from the switch is nonproprietary (as with analog and ISDN lines).

7.5.8. Other Implementation-Specific Ports

A variation of the proprietary add-in board configuration involves the use of a parallel port or some other implementation-specific computer port to attach a CTI component. Like the add-in board, this type of CTI component must be accompanied by a mapper of some sort, which at a minimum will allow access to the CTI stream provided by the component.

7.6. Client-Server Configurations

Client-server configurations, as the name implies, involve an indirect communication path between a user's client computer or PDA and a telephone station or other telephony resources.

All of the configurations in this section are presented using the standard graphical notation. Refer to the inside of the back cover for a summary of the symbols.

7.6.1. Basic Client-Server Configuration

In the simplest of *client-server configurations*, a client computer or PDA establishes a communication path with a CTI server which acts as a proxy in obtaining CTI functionality from a switch.

Figure 7-33 shows the *logical integration* of a client computer and a telephone station in a client-server configuration involving a single client computer. From the perspective of a user controlling the functionality of their telephone station, the indirect flow of CTI messages through the CTI server is functionally equivalent to the direct flow of messages found in direct-connect configurations (assuming that the CTI interfaces themselves have the same functionality).

In the configuration shown, both the switch and the CTI server provide standard CTI protocols, so no proprietary software is required on either the CTI server or the client computer.

Figure 7-33. Client-server configuration example

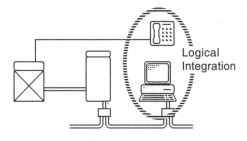

Logical Integration

If the telephony features and services available through the resulting CTI stream reflect a switching domain that is limited in scope to a single device or device configuration, then the configuration involves first-party call control. This is illustrated in Figure 7-34. If the switching domain contains additional telephony resources, then it supports third-party call control. The switching domain in this case may consist of other station devices and/or additional telephony resources within the switch (Figure 7-35). In either case, CTI messages from the switch are appropriately delivered through a LAN-based communication path between the CTI server and the client computer or PDA.

Figure 7-34. Client-server first-party call control

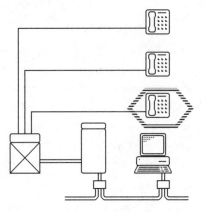

One of the most significant benefits of client-server CTI system configurations involves taking advantage of the fan-out capability of a given CTI server. By making a given CTI server available on a LAN, a client-server configuration can be easily scaled to any size and can include any type of PDA, personal computer, multi-user computer, or other logical client. This is illustrated in Figure 7-36. The CTI server in the configuration shown supports standard CTI protocols so any CTI Plug & Play hardware component can establish a communication path with it (after using the appropriate user ID and password if necessary).

Figure 7-35. Client-server third-party call control

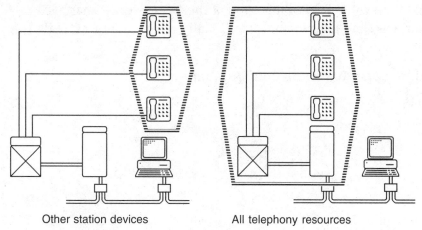

Other station devices   All telephony resources

Figure 7-36. Client-server LAN configuration

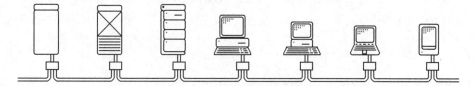

7.6.2. Client-Server Mapper Configurations

Three different kinds of mappers may be found in client-server CTI system configurations because they encompass both switch-to-CTI server links and CTI server-to-client links:

- Switch-server mapper hardware

- Switch-server mapper code

- Server-client mapper code

Switch-Server Mapper Hardware

The role of a switch-server mapper hardware component is illustrated in Figure 7-37. In this example, the mapper hardware (labeled "P") allows the switch, which supports only a proprietary CTI protocol, to communicate with the CTI Plug & Play CTI server, which supports CTI Protocol 1. The combination of the switch-server mapper hardware and the switch is functionally equivalent to a CTI Plug & Play switch.

Figure 7-37. Mapper hardware configuration

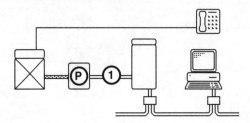

Switch-Server Mapper Code

Figure 7-38 shows a client-server configuration in which the proprietary mapper code, written and provided by the switch vendor, is installed on the CTI server so that it can encode and decode the proprietary CTI protocol provided by the switch.

Figure 7-38. Switch-server mapper code configuration

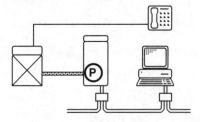

Server-Client Mapper Code

In the system configuration depicted in Figure 7-39, the CTI server uses a proprietary protocol across the LAN communication path and thus does not support integration with CTI Plug & Play components unless proprietary mapper software is installed on each client computer, PDA, etc., on the LAN. Figure 7-40 illustrates the magnitude of the challenge associated with supporting server-client mapper code for a large network of diverse products. If a CTI server does not support a standard CTI protocol, and is therefore not CTI Plug & Play, mapper software may or may not be available for every type of product on a given network.

Figure 7-39. Server-client mapper code configuration

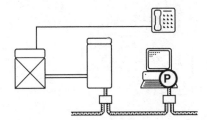

Figure 7-40. Server-client mapper code LAN configuration

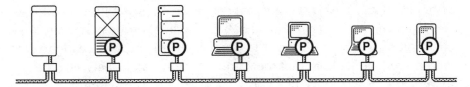

7.6.3. Client-Server Media Stream Access Configurations

When media access service requests are used to bind a media service instance to a call, a media access device (which may be a station device as far as the switch is concerned) that is associated with the appropriate media server is connected to the call in question. The media access device may be added (resulting in a multi-point call) or it may take the place of a device previously participating in the call.

CTI Plug & Play Media Access

Figures 7-41, 7-42, and 7-43 present three different CTI system configurations in which media access services are supported using standard CTI protocols and well-defined media service protocols. The CTI streams and media service streams flow between the CTI servers and their logical clients without requiring proprietary software, making these configurations CTI Plug & Play.

The system configuration shown in Figure 7-41 features a hybrid product consisting of a CTI-server and a media server. The media service instance and the CTI server implementation are integrated on a single component so all binding of media stream identifiers is simplified.

Figure 7-41. CTI server with media access resources

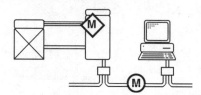

The system configurations shown in Figures 7-42, and 7-43 involve distinct CTI and media servers that cooperate to present the client computer shown with a media-capable switching domain. In each case one server front-ends the other to provide the media service identifier binding function. (Note that the communication paths between the servers are shown above one another to better illustrate their relationship. Typically both servers would be connected to the same LAN.)

Figure 7-42. CTI server front-ending media server

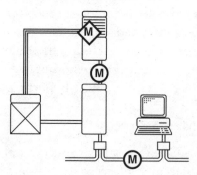

Figure 7-43. Media server front-ending CTI server

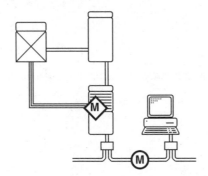

Server Mapper Media Access

Figure 7-44 depicts the use of media server mapper code (as defined in section 7.4). In this system configuration, there is no direct interaction between the CTI server and the media server. It is the responsibility of the special media server mapper code installed on the client computer (labeled "S") to provide the media binding functionality. The fact that the CTI server in this configuration supports CTI Plug & Play means that in this case mappers are only required to integrate the media services with the standard CTI protocol provided by the CTI server. Media access services can then be used by other software components running on the client computer.

Figure 7-44. Media service mapper code

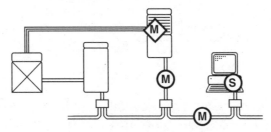

In Figure 7-45, a variation on the previous system configuration is shown. In this case, the CTI protocol does not support CTI Plug & Play and requires that its logical clients use mapper code to work with its proprietary protocol. In this case, the same media service mapper (labeled "S") from the previous example continues to work but it is actually layered over the mapper code for the CTI server (labeled "P"). The CTI server's mapper code delivers a standard CTI protocol which can, in turn, be used by the media server mapper.

Figure 7-45. Layered media service mapper code and CTI mapper code

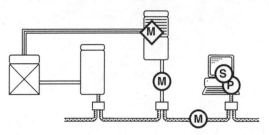

7.6.4. LAN Dial-Up Bridge Configuration

Dial-up bridges allow hardware components to remotely connect to a LAN. Figure 7-46 illustrates a variation on the basic client-server system configuration depicted in Figure 7-33 in which the personal computer is located in a remote location from the telephony resources being observed and controlled.

Figure 7-46. LAN remote access configuration

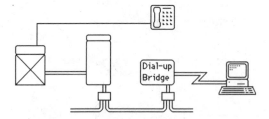

This configuration is useful for remotely monitoring a particular telephone station or telephony resources such as ACD groups. For example, an employee working at home could monitor calls being routed to her office phone and selectively transfer them to her home.

7.6.5. LAN Dial-Up Bridge Configuration/OPX

This configuration is a variation on the basic dial-up LAN remote-access configuration described above. The extension being monitored in this case is an off-premises extension (OPX) that is co-located with the remote CTI client. This is illustrated in Figure 7-47.

Figure 7-47. LAN remote access configuration with OPX

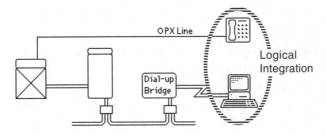

This configuration is the standard approach for work-at-home call center agents. With this arrangement, any user can have all the telephony functionality he or she would have in the workplace.

7.6.6. LAN Dial-Up Bridge Configuration/SVD

This configuration is yet another variation on the LAN remote-access configuration. The dial-up bridge in this case utilizes simultaneous voice and data (SVD) technology to provide access to the voice channel associated with the given user's phone line on the PBX, despite the remote location. This functionality, shown in Figure 7-48, is equivalent to the OPX line approach but requires only a single media stream channel (or analog phone line).

Figure 7-48. LAN remote access configuration with SVD

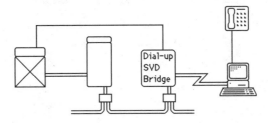

7.6.7. CO-Server Dial-Up

The ability to provide CTI capabilities to remote users is not restricted to users of CPE switches. Local exchange carriers may also make CTI access available to their CO switches. Figure 7-49 shows a CTI configuration in which client computers use a dial-up communication path to reach a LAN on which the CTI server is located. This LAN could be a private LAN managed by the carrier or it could be a segment of the public Internet. Figure 7-50 shows a simplification of this approach where the CTI server itself supports dial-up communication paths and no LAN is required. Both these configurations utilize dual media stream channels (e.g., two analog lines, two ISDN channels) at the user's location. For this reason CO-server dial-up is optimal for a multi-channel line such as ISDN BRI or cable TV.

Figure 7-49. CO-server remote access configuration using dial-up bridge

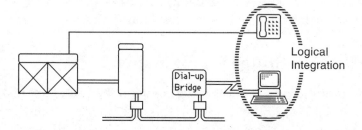

Figure 7-50. CO-server remote access configuration using dial-up server

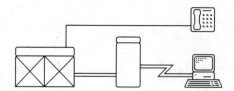

CTI services provided by the local exchange carrier are already available for certain Centrex customers, but they represent a very attractive offering for home businesses and telecommuters. Local exchange carriers acting as Internet service providers can combine this with their offerings.

7.6.8. CO-Server Remote Access/SVD

SVD technology can be used to overcome the need for two phone lines in CO-switch CTI configurations. SVD technology may be incorporated into a dial-in bridge (Figure 7-51) or into a specialized CTI-server that provides support for SVD dial-up (Figure 7-52).

Figure 7-51. CO-server remote access configuration using SVD dial-up bridge

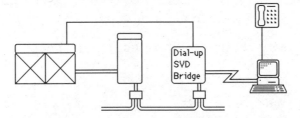

Figure 7-52. CO-server remote access configuration using SVD dial-up server

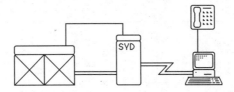

7.7. Client-Client Configurations

A common scenario in many office environments involves using a pair of products, such as a personal computer and a laptop, or a personal computer and a PDA, only one of which can be directly integrated into a supported CTI configuration. Client-client configurations involve using a special software component installed in one of the products, perhaps the personal computer, to let the other product act as a secondary client.

Figures 7-53 and 7-54 depict two typical client-client system configurations. In both, a PDA uses an infrared communication path to send CTI messages to a personal computer. The personal computer acts as a proxy in what are otherwise basic direct-connect and client-server configurations. Any of the system configurations presented in this chapter can be extended in this fashion.

Figure 7-53. Client-client configuration, direct-connect case

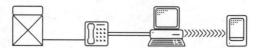

Figure 7-54. Client-client configuration, client-server case

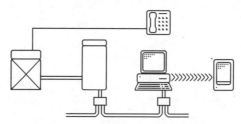

7.8. Review

In this chapter we have seen that a limitless variety of CTI system configurations can be assembled from a set of basic *CTI hardware components,* given standard *communication paths* and *CTI protocols.*

CTI hardware components include *client computers* (*personal* and *multi-user*), *personal digital assistants* (*PDAs*), *telephone stations, telephone station peripherals, CTI servers, media servers, switches* (*CPE* and *CO*), *Internet voice gateways,* and *hybrids* of all of these basic types.

Communication paths, which include both the physical link layer and the complete session/transport protocol stack, are broadly categorized as *local area network* (*LAN*), *serial cable* and *serial bus, infrared,* and *dial-up.*

The communication paths between adjacent CTI hardware components in a CTI system are used to send a *CTI stream* using a *CTI protocol,* and they may or may not carry one or more *media service streams* for any media services that are accessed. CTI protocols are either *standard* or *proprietary.* Standard CTI protocols allow for CTI Plug & Play operation.

Protocol mappers are components that translate between a proprietary CTI protocol and a standard CTI protocol in order to let a hardware component that relies on proprietary protocols interoperate with CTI Plug & Play hardware components. They may be implemented as *mapper hardware* that can connect within a communication path and will work with any hardware components, or they can be implemented as *mapper code* that must be installed and run on every hardware component configured as a logical client, and that must be developed for each platform individually.

CTI hardware components that implement proprietary protocols may provide *API-specific adapter software* for specific operating systems that offer programmatic CTI interfaces (APIs). Names for adapter software vary depending on the operating system, but include *driver, service provider,* and *telephone tool.* Adapter software may be developed to support CTI system configurations described in this chapter, but these can be used only in the configuration(s) and on the operating system(s) for which they were implemented.

CTI hardware components that implement proprietary protocols also may be found within turnkey *custom CTI solutions*. These solutions may be developed to encompass one or more CTI system configurations described in this chapter, but once again can be used only in the configuration(s) and on the operating system(s) for which they were implemented.

8.
CTI Software Components

In this chapter we explore the software components that go into building a CTI system. There are many different types of software components that may play a role in a CTI system. These range from protocol stack implementations, to various APIs, to different sorts of application software. The chapter covers the layering and relationships among the various types of CTI software components, and describes the role of each type of component in a CTI solution.

While the overall framework for CTI software components is independent of the operating system involved, the implementations of certain components represent significant de facto standards in their own right. Sections of this chapter therefore specifically describe certain portions of the CTI software framework as they are implemented for the two leading personal computing operating systems: Apple Computer's Mac™ OS and Microsoft's Windows™ operating system. Apple's framework for CTI is called the *Macintosh Telephony Architecture* or *MTA* for short. Microsoft's set of specifications is called *Windows Telephony*.

8.1. CTI Software Component Hierarchy

Almost every imaginable CTI hardware component is based on computer technology and thus is managed by software that is responsible for sending, receiving, interpreting, generating, and handling CTI messages. In assembling CTI systems and building individual CTI components, however, we are concerned only with software environments designed for the installation or addition of software[8-1] developed by third parties. Within such an environment, CTI software components are arranged in a functional hierarchy—a *CTI software framework*, that corresponds to the CTI value chain.

8.1.1. CTI Value Chain

The CTI value chain (introduced in Chapter 2) represents the various specialized groups that contribute components to any given CTI system. At this point in our exploration of the technologies that support this value chain, we have looked primarily at the lower levels of the pyramid, where the telephony resources are located. In Chapter 7 we explored how these are embodied and physically connected to form systems of hardware components. In this chapter we turn to the upper layers of the value chain, where software vendors and system integrators add their value to solutions.

In these tiers of the value chain we are concerned primarily with:

- Operating system and/or network operating system software components that support CTI

- Products from telephony software developers

8-1. Closed CTI hardware components — CTI hardware components that are not open to the addition of software components range from switches to PDAs. Internally any of these components may in fact be built using off-the-shelf computer hardware, operating systems, and standards-based telephony resource components (such as S.100 components). The internal architecture of these devices (and related standards) are beyond the scope of this book, however, because they are not of direct consequence to a CTI system.

- Mainstream applications

- Scripts, utilities, and other software developed by systems integrators, customers, and individuals customizing their systems

Historically, CTI integration has taken place exclusively in the realm of software, without consideration for the plug-and-play interoperability of hardware components. For this reason, the generalized software framework also provides for software components that are provided by individual telephone equipment vendors.

8.1.2. Modularity

A CTI system consists of a collection of individual CTI components in which each communicates with its neighbor(s) using CTI messages that travel across an inter-component boundary (as described in Chapter 6).

Modularity is just as important for software components as it is for hardware components. System integrators, customers, and individuals want to be able to plug and play with their software, just as they can plug and play with their hardware. This is fundamental to the concept of the CTI value chain.

For CTI hardware components, the inter-component boundaries are in the form of CTI protocols that travel over communication paths between hardware components (as illustrated in Chapter 7). For software components they are in the form of programmatic interfaces through which CTI messages are communicated.

8.1.3. Programmatic Interfaces

Just as hardware components use transmitters, receivers, cables, and connectors to support the communication paths that carry CTI streams, two distinct software components running on the same hardware component can link with one another and communicate using a *programmatic interface*.

Figure 8-1. Programmatic interfaces

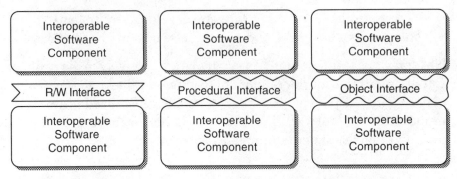

Programmatic interfaces are software-based inter-component boundaries as shown in Figure 8-1. Software components interoperating through a programmatic interface boundary are typically arranged vertically in diagrams to reinforce the fact that both software components exchanging messages through a particular programmatic interface are on the same computer.

Three distinct categories of programmatic interfaces are of concern to CTI systems:

- Read/write interfaces

> R/W Interface

Read/write interfaces, or just R/W *interfaces* for short, are simple programmatic interfaces that allow a software component to obtain access to a *protocol stack implementation* (or a communications driver) in order to open or close a communication path to another software component (on the same or a different hardware component), and to read or write a stream of data across the communication path. If the stream is for carrying CTI messages, it is referred to as a *CTI stream* and the messages are delivered in the form of encoded CTI *protocol data units* (*PDUs*). R/W interfaces and protocol stack implementations are described in further detail in the next section and in section 8.3.1.

• Procedural Interfaces

Procedural interfaces are frequently referred to as Application Programming Interfaces or APIs.[8-2] Procedural interfaces allow two software components to communicate through a set of well-defined function calls. In contrast with a R/W interface, a procedural interface often uses many different function calls for exchanging messages. Message parameters are placed in structures or simple variables and passed across the interface through references in function parameters.

• Object Interfaces

Object interfaces allow a software component to access telephony functionality by manipulating software objects. Unlike procedural interfaces, formal object classes are used to define the programmatic interface and the fashion in which information is exchanged.

8.1.4. CTI Software Framework

The CTI software framework that forms the basis for integrating CTI software components on a given hardware component is shown in Figure 8-2.

8-2. API — In general usage the term *API* (which technically stands for *Application Programming Interface*) refers to any programmatic interface that allows two independent software components to interact. It is not limited to application software programming. To avoid any confusion, however, this book uses the term *programmatic interface* to refer to interfaces between software components. The term *SPI* (for *Service Provider Interface*) and other acronyms are used by some vendors to refer to programmatic interfaces that are not intended for application developers.

Figure 8-2. CTI software component framework

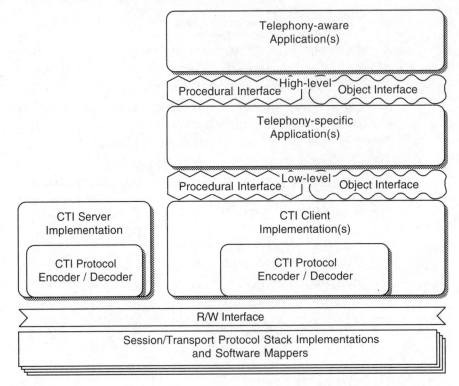

Software component types are listed below, starting from the bottom and working up.

- Session/transport protocol stack implementations

    Session/transport protocol stack implementations and associated communications drivers are usually provided either by operating system vendors or network operating system vendors. In cases where the desired CTI stream must be accessed through a piece of proprietary hardware, or a proprietary protocol stack, one or more of these components must be provided by the vendor of the logical CTI server.

- Software mappers

    Mappers (introduced in Chapter 7, section 7.3.2) translate between proprietary CTI protocols and standard CTI protocols. They are implemented as session/transport protocol modules so that they are transparent to all of the other software components. These are, of course, provided by the vendor of the logical CTI server to be used in a given instance.

- R/W interface

    The *R/W interface* on a particular platform typically is provided by the operating system vendor.

- CTI server implementations

    CTI server implementations typically are provided by telephony software developers and network operating system vendors. Client-client configurations (as described in Chapter 7, section 7.7) are a special case in which simple CTI server implementations may be provided by operating system vendors.

- CTI client implementations

    CTI client implementations are software components that manage the communication path to a particular logical CTI server; they translate the CTI protocol that flows through the resulting CTI stream for access through a low-level procedural CTI interface, a low-level object CTI interface, or both. There are two types of these software components. *CTI Plug & Play client implementations* support standard CTI protocols, which flow across communication paths either from connected logical servers or from software mappers, installed below the R/W interface, that translate from a proprietary protocol. They are provided by operating system vendors or telephony software developers. On the other hand, *API-specific adapters* are provided by an individual telephony equipment vendor for use with the proprietary protocol or driver associated with their product. They are called API-specific because each is written to support a specific programmatic interface (typically called an API) on a specific operating system.

- Low-level procedural and low-level object interfaces

  - *Low-level procedural CTI interfaces* and their object-based counterparts, *low-level object CTI interfaces*, are programmatic interfaces that represent the service boundary used by telephony-specific applications to access the switching domain for a particular CTI system. The three most popular low-level CTI interfaces are TAPI, Telephone Manager, and TSAPI.[8-3]

  - *Procedural media service interfaces* and their object-based counterparts, *object media service interfaces*, are programmatic interfaces that provide applications with access to media services such as data communications, audio recording and playback, speech, and video. These interfaces are the standard media service clients that are embedded within every operating system, and they complement the corresponding CTI interfaces by providing the ability to manipulate the media streams associated with calls using existing media software on a given operating system platform.

- Telephony-specific applications

  Telephony-specific applications are software components with observation (and optionally control) of telephony resources as their primary mission. Telephony-specific applications can be categorized as either being either *screen-based telephone* applications or *programmed telephony applications*.

8-3. Popular CTI interfaces — TAPI stands for *Telephony Application Programming Interface*. It is provided by Microsoft as part of the Windows operating system. Telephone Manager is provided by Apple Computer as part of the Mac OS. TSAPI stands for *Telephony Services Application Programming Interface*. The definition of TSAPI is in the public domain, so any operating system vendor, network operating system vendor, or telephony software developer is free to implement TSAPI. See section 8.4 for details on each of these procedural interfaces.

- High-level procedural and high-level object interfaces

  High-level procedural CTI interfaces and their object-based counterparts, *high-level object CTI interfaces*, are programmatic interfaces designed to support telephony-aware applications. They are optimized to allow mainstream application developers, system integrators, and end users to easily add telephony support to their products, solutions, and work environments. While the functions associated with high-level CTI interfaces generally are implemented and defined by operating system vendors, these interfaces are actually used to access functionality in a designated telephony-specific application. For this reason they are layered above telephony-specific applications, and they represent the service boundary between telephony-aware applications and telephony-specific applications.

- Telephony-aware applications

  Telephony-aware applications (also called *telephony-enabled applications*) are, by definition, any application or other piece of software (such as utility or script) taking advantage of one of the high-level CTI interfaces. Despite the fact that they may be very much removed from the core of telephony functionality and telephony resources, they represent the primary vehicle through which system integrators, customers, and individuals see much of the immediate benefit of telephony solutions. As a result of the ease with which telephony-aware applications can be written (in marked contrast with every other layer in the software framework), the number of available telephony-aware applications will be many orders of magnitude greater than telephony-specific applications. However, each plays an important role in a complete CTI solution.

An instance of each of these CTI software component types may or may not exist in any given CTI system implementation. For any particular software component to actually work, however, all of the components it depends on must be in place below it.

It is likely for a complete CTI solution to involve many different applications, and for there to be only one (or fewer) of each of the other types of software components. However, it is possible to have any number of each of these component types running on the same hardware platform.

8.2. CTI Server Implementations

CTI server implementations are the actual software components that correspond to the CTI server hardware components described in Chapter 7. A CTI server implementation may play one or more of the following roles in a CTI system:

- It acts as a fan-out component (described in Chapter 6, section 6.3.3)

- It acts as a security firewall between logical clients and an unsecured CTI interface (such as one found on a switch).

- It translates among different protocols, allowing logical clients using one protocol to communicate with a logical server using a different protocol.

- It augments the functionality of its logical server to provide a richer set of telephony features and services to its logical client.

A generic CTI server implementation operates as follows:

- It is configured to connect to one or more logical servers (often directly to the CTI interface of a switch), and it maintains these communication paths over time.

- It is configured to accept communication path requests from one or more (typically many) logical clients using one or more session/transport protocol stacks. If the CTI server implementation supports security features, the protocol stacks used must support authentication. Authentication may be as simple as layering an initial user ID and password exchange mechanism on top of a non-authenticating protocol, or it may involve a more sophisticated mechanism such as the exchange of encrypted keys as implemented in such protocols as Secure Sockets Layer (SSL).[8-4] The authenticated protocol may or may not also employ encryption for complete protection of all information traveling across the CTI stream.

- If the CTI server supports security, it also is likely to support user privilege restrictions. This involves linking an individual's authentication identity to a database of privileges. Depending on the implementation, privilege information for each user may include:

  - What devices in the switching domain are visible. For example, a secretary may be permitted to see only her phone and her manager's phone.

  - What services may be requested for each device that is visible. The secretary above might have access to all services for her own phone, but only the *start monitor* and *stop monitor* services for her manager's phone.

- When a logical client (normally a CTI client implementation, but possibly another downstream CTI server implementation) attempts to establish a communication path to the CTI server implementation, the attempt is first authenticated if security is being used. The pair of communicating components then exchange protocol and version negotiation packets to determine whether or not they can interoperate and, if so, which protocol to use. Assuming they agree on a protocol, and if the server indicates to the client that it is in an operational state, the logical client completes the start-up process by requesting capabilities information from the CTI server implementation. The CTI server implementation provides the appropriate capabilities information based on:

  - The capabilities of its logical server;

  - Any capabilities that it adds; and

  - Any privilege restrictions that apply to the logical client.

8-4. Secure Sockets Layer — Secure sockets layer is a protocol for secure communication over the Internet.

- At this point, the CTI stream between the CTI server implementation and its new logical client continues until the logical client shuts down the communication path or the communication path fails for some reason. The CTI server is responsible for acting as the full proxy for the logical client. This includes, for example:

  - Ensuring that for every service request the logical client issues, it receives the appropriate positive or negative acknowledgment;

  - Ensuring that every event the CTI server implementation receives (from its logical server) that applies to one of its logical client's active monitors is delivered to the logical client with the correct monitor cross-reference identifier.

CTI server implementations support CTI Plug & Play, or use proprietary interfaces, or may be a combination of the two.

8.2.1. CTI Plug & Play Servers

A *CTI Plug & Play server implementation* is one that is based on standard CTI protocols. It uses the appropriate R/W interfaces for the platform on which it is running, and connects with both logical servers and logical clients using standard protocols over these communication paths. This is illustrated in Figure 8-3.

A logical server that delivers a proprietary protocol across the communication path may provide a software mapper in order to interoperate with CTI Plug & Play server implementations. This is illustrated in Figure 8-4.

A CTI Plug & Play server implementation generally supports two or all three of the standard CTI protocols for communicating with both logical servers and logical clients. At a minimum, it will support CTI Protocol 1 for the CTI stream with its logical server (typically a switch) and CTI Protocol 2 for the CTI stream with the logical clients. It then translates these protocols (which process involves only minimal changes to the encoding) between one boundary and the other.

Figure 8-3. CTI Plug & Play server implementation

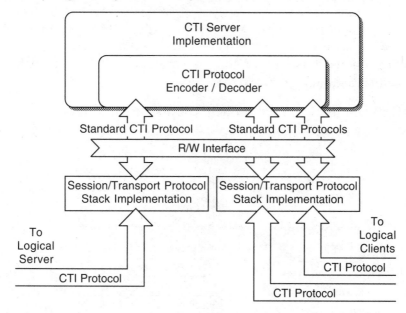

Figure 8-4. CTI Plug & Play server implementation with mapper

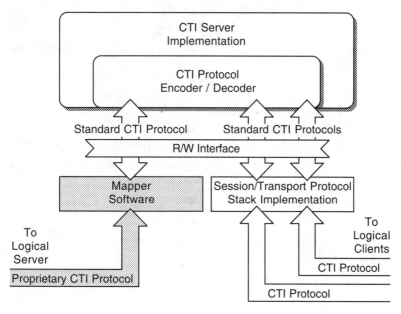

8.2.2. Proprietary Interface Servers

A CTI server implementation with proprietary interfaces assumes that its logical server is a switch and its logical clients are client implementations; it uses proprietary interfaces to connect with each. This is illustrated in Figure 8-5.

Figure 8-5. CTI server implementation with proprietary interfaces

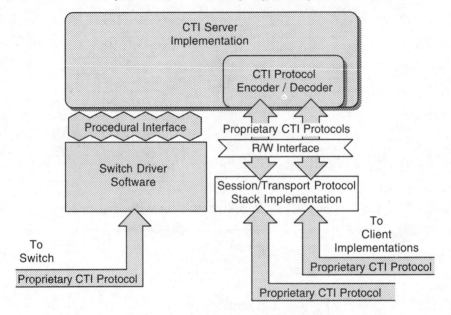

The CTI server implementation vendor defines a procedural API for connecting to the switch using special switch driver software, a form of API-specific adapter software for server implementations. The switch driver software may be written by the switch vendor or by the vendor of the CTI server implementation. Proprietary protocols are then fanned out to corresponding CTI client implementations.

8.3. CTI Client Implementation

The *CTI client implementation* is the software component in a CTI system that is responsible for ensuring that a particular logical CTI server can be accessed through one or more programmatic interfaces.

8.3.1. R/W Interfaces

The programmatic interface used to access session/transport protocol stacks (across which CTI protocols are passed) is called a *R/W programmatic interface*, or just *R/W interface* for short. A R/W interface allows software components to open and close communication paths and to read and write the corresponding streams of data using session/transport protocol stack implementations.

Features of a R/W interface include:

- Activating (open) and deactivating (close) session/transport protocol stack implementations;

- Layering session/transport protocol stacks on top of one another;

- Supporting multiple streams and multiple protocol stacks simultaneously;

- Sending (write) and receiving (read) to and from streams; and

- Translating a reference for a desired communication path destination into the correct protocol stack and destination address.

CTI messages travel through a R/W interface in the form of buffers of encoded (and thus self-contained) PDUs (Figure 8-6). The R/W interface lies between a CTI protocol encoder/decoder and a session/transport protocol stack implementation instance managing an active CTI stream.

Software components, typically CTI client implementations and CTI server implementations, use R/W interfaces to establish communication paths among hardware components and to exchange

Figure 8-6. R/W interface

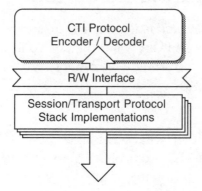

CTI messages as encoded PDU information. Unlike procedural and object CTI programmatic interfaces, R/W interfaces are general purpose mechanisms designed to allow manipulation of session/transport protocol stacks. In contrast to the other programmatic interface types, using the R/W interface itself is quite simple and any complexity comes only in the process of encoding and decoding the PDUs.

Common session/transport protocol stacks that may be accessed through R/W interfaces (as applicable to a given computer system) include the following, as shown in Figure 8-7:

- ADSP

- Infrared

- IPX/SPX

- RS-232 and virtual serial ports

- TCP/IP

A very important feature of R/W interfaces is the ability to layer one protocol stack implementation on top of another. Figure 8-8 shows an example where a TCP/IP protocol stack implementation has used PPP to establish a remote IP session over a modem protocol stack which is communicating with modem hardware using an RS-232 driver.

Figure 8-7. Session/transport protocol stack implementations

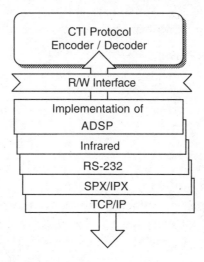

Figure 8-8. Example of layering with R/W interfaces

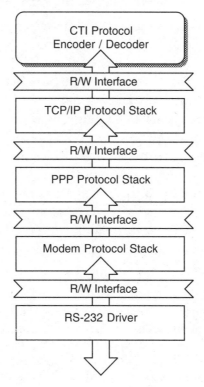

Mainstream operating systems support standard R/W interfaces for managing and using session/transport protocol stack implementations and communication drivers. Examples include:

- Macintosh: Open Transport[8-5]

- Windows: Windows Sockets[8-5.]

- Unix: Streams

8.3.2. Software Mappers

Software mappers are software components that are constructed as session/transport protocol stack implementations for the R/W interface on every platform. The role of a software mapper is to translate between a proprietary CTI protocol and a standard CTI protocol in order to support transparently CTI client implementations (and CTI server implementations) that are CTI Plug & Play.

Software mappers appear to the R/W interface just like any other session/transport protocol stack. They are therefore used just like any other protocol stack implementation. When a CTI client implementation wants to connect to a logical server that uses a mapper, it simply opens a communication path as it normally would, but it specifies the mapper as the protocol stack to use. Figure 8-9 provides two examples of a software mapper in operation. In both cases the Phones-R-Us logical server is accessed using the Phones-R-Us software mapper. In the case on the left, the Phones-R-Us logical server happens to use the TCP/IP protocol stack to transport its proprietary protocol, and the Phones-R-Us mapper layers itself on top of this protocol stack and generates a standard CTI protocol. In the case on the right, the CTI logical server is actually an add-in board that is accessed through a driver.

8-5. Standard R/W interfaces — Secondary R/W interfaces for Mac OS include the Connection Manager and the serial port interface. Secondary R/W interfaces for Windows include the file and communications functions. CTI client implementations (and CTI server implementations) may support as many R/W interfaces as they wish, but they must at least support the primary R/W interfaces listed in order to support software mappers.

Figure 8-9. Software mapper example

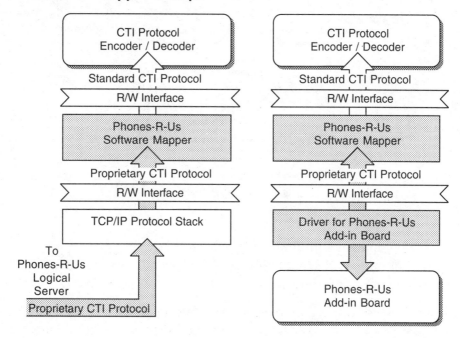

8.3.3. CTI Plug & Play Client Implementations

A *CTI Plug & Play client implementation* is a CTI software component that both supports low-level programmatic CTI interfaces (one or more) and contains a CTI protocol encoder/decoder in order to support standard CTI protocols (one or more) as shown in Figure 8-10.

CTI Plug & Play client implementations are developed specifically for a given operating system platform. They support the programmatic interfaces (both CTI and media) appropriate for that operating system.

Support for low-level procedural interfaces involves mapping the CTI messages traveling across the R/W interface in the form of PDUs into a collection of procedural function calls. Support for an object interface involves doing a similar mapping operation using formal object classes that correspond to the CTI abstraction. In either case, this also involves making sure that multiple applications using any of the exposed interfaces are able to simultaneously access the same

Figure 8-10. CTI Plug & Play client implementation

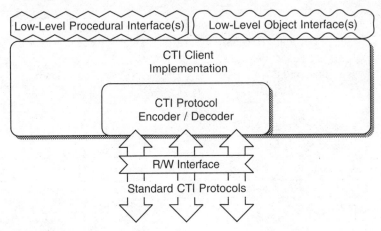

switching domain without having to establish multiple communication paths. In other words, CTI client implementations are software fan-out components.

In addition to supporting low-level CTI interfaces by appropriately mapping the contents of CTI messages, the CTI Plug & Play client implementation also is responsible for translating references to media services to the appropriate programmatic interface for a given media type on the operating system in question.

 As with every other CTI component in a CTI system, the CTI Plug & Play client implementation software can incrementally add value to the switching domains that it is representing to logical clients across the programmatic interfaces. For example, if a given CTI client implementation determines that a particular switching domain does not have support for canonical phone number format, and the CTI client implementation itself is able to perform the necessary translation, it can report to its logical clients that their switching domain does support canonical numbers.

8.3.4. API-Specific Adapter Software

API-specific adapters are CTI client implementations that support a single platform-specific programmatic interface and interact directly with the CTI resources they represent, typically through a proprietary CTI protocol of some sort. Because these pieces of software are API-specific, they need to be described in the context of the APIs with which they are associated.

Windows Telephony (TAPI): Telephony Service Providers

Windows Telephony defines the components used to provide access to a particular source of CTI functionality to clients of the TAPI interface as *telephony service providers*. They are software components that are implemented as Windows *dynamic link libraries* (*DLLs*). In this model, a complete CTI client implementation typically consists of the telephony service provider and some associated drivers that interact with the CTI logical server, and may in fact be made up of several different modules, including media drivers and proprietary hardware drivers.

Microsoft's specifications for Windows Telephony define a procedural interface called the *telephony service provider interface*, or *TSPI* for short, that represents the boundary between the TAPI implementation and the telephony service provider.[8-6] This is illustrated in Figure 8-11.

Macintosh Telephony Architecture (Telephone Manager): Telephone Tools

The low-level portion of the Macintosh Telephony Architecture is based on the Telephone Manager. The Telephone Manager defines the vendor-specific software components it uses to access CTI functionality as *telephone tools*. In the Macintosh Telephony Architecture, telephone tools are coupled with a *device handler* and one or more media drivers. This is illustrated in Figure 8-12.

8-6. Telephony Service Provider Interface — The Telephony Service Provider Interface, or TSPI, is documented in the book *Telephony Service Provider Programmer's Reference* available from Microsoft through the Microsoft Developer Network Library.

Figure 8-11. Windows telephony service provider

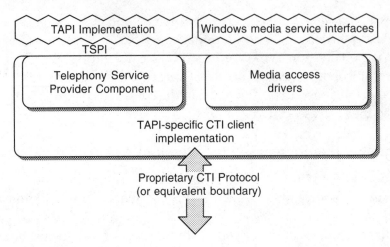

Figure 8-12. Mac OS telephone tools

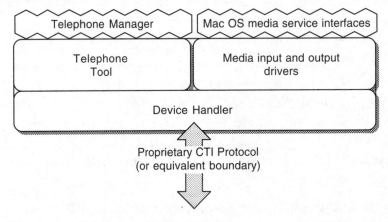

The device handler portion is responsible for handling the media streams associated with a given CTI service, and for ensuring that media access is shared and appropriately reflected by both the media drivers and the telephone tool. The structure for the device handler is not defined by Apple.[8-7]

8.4. CTI Low-Level Application Programming Interfaces (APIs)

Programmatic interfaces for CTI fall into three categories:

- Implementation-specific

    Implementation-specific CTI interfaces are those in which a proprietary interface is defined and implemented as part of a particular CTI client implementation.

- Operating-system-specific

    Operating-system-specific CTI interfaces are defined and implemented as part of an operating system implementation. The two most popular personal computing operating systems, Mac OS and Windows, each have their own low-level APIs for CTI. These are the Telephone Manager and TAPI respectively.

- Platform-independent

    Platform-independent CTI interfaces are those that are designed to work with any platform. TSAPI[8-8] is the only platform-independent low-level CTI interface.

TSAPI is characterized as a very thin API layer that provides direct access to standard CTI protocols. TSAPI's primary goal is ensuring that every piece of available information and every accessible capability of a given switching domain implementation is made available to client software, independent of operating system platform. The specification of the TSAPI interface is focused strictly on

---

8-7. Telephone Tool specifications — Documentation for developing telephone tools is contained in the book *Telephone Manager Developer's Guide* that is part of the Mac OS Software Developer's Kit. This product is available from Apple through APDA. (*Telephone Manager Developer's Guide*, Apple Computer, Inc., 1991.)

8-8. TSAPI — All references to TSAPI are to the definition of Versit TSAPI as defined in the *Versit Computer Telephony Integration (CTI) Encyclopedia*. Earlier versions of TSAPI, including Novell's TSAPI implementation for the NetWare environment, are effectively subsets of Versit TSAPI.

accessing and using the CTI functionality represented by the standard CTI protocols. It provides a layer above the raw protocol that simplifies software access by providing functionality for encoding and decoding and managing the queues of incoming and outgoing messages, without providing any interpretation of the protocol that might lead to loss of information or accuracy.

On the other hand, OS-specific interfaces were designed and developed with slightly different goals. The principal focus of these APIs is to simplify the programming task for the majority of CTI application developers who are using the facilities of the operating system in question. This has resulted in simplifications that may help most application developers, but also has reduced the information available for certain types of applications. In other words, operating system interfaces attempt to provide an additional layer of functionality above the raw messages from the switching domain. These extras include:

- Providing a simplified first-party call control model;

- Presenting distinct interface subsets for the three different aspects of CTI (call control, telephone control, media access);

- Tracking switching domain capabilities;

- Tracking of call identifiers on behalf of applications;

- Mapping switching domain identifiers into handles and local identifiers;

- Tracking state, status, and setting information on behalf of applications;

- Tracking what services are and are not applicable to a connection at a given instant;

- Tracking service completion on behalf of applications;

- Abstracting differences between digital data and voice calls;

- Supplying mechanisms for arbitration of call ownership between applications; and

- Integrating tightly with OS-specific media service APIs.

Of all theses features, the first is the most significant distinction. Existing operating-system-specific interfaces (TAPI and the Telephone Manager) derive much of the simplification they offer to application developers by abstracting all switching domains as being first-party. (Refer to Chapter 6, section 6.5.2 for the definition of first-party call control.) This simplification reflects the fact that most CTI applications are concerned only with a single device. Depending on the service provider or telephone tool being used, support for logical CTI servers offering third-party call control is accomplished either by representing every device as a distinct switching domain, or by representing it as many devices that are part of a large device configuration.

In practice, either type of low-level API can be used for most applications, assuming that any needed functionality factored out by the OS-specific API still can be accessed through an escape mechanism in that API. When developing a telephony-specific application, the following trade-offs must be considered in choosing between using a platform-independent interface (TSAPI) or an OS-specific interface (TAPI and the Telephone Manager):

- First-party call control
    If an application only requires the ability to observe a single device or device configuration, either type of API may be used. If an application needs the ability to track calls that travel between multiple devices, it probably will be easier to develop using TSAPI.

- Application short cuts

    TAPI and the Telephone Manager provide interfaces in which much of the work associated with tracking information from the switching domain is handled automatically. When the information is needed by an application, it is immediately available. On the other hand, applications using TSAPI must themselves track all of the pertinent pieces of information that arrive in messages and might be needed later. This includes tracking information to determine what service requests can be applied to a given call given its state, and the presence of dynamic feature presentation information if provided.

- Veracity

    TSAPI provides the most complete view of what is taking place in the switching domain. While the functionality of the other APIs is quite good, they do involve many more simplifications (and hence mappings) than TSAPI, and ultimately they represent a subset of the full feature set that TSAPI has access to through standard CTI protocols.

- Portability

    While implementations of TSAPI on different platforms must differ slightly simply to adapt to the way each operating system works, the main benefit of a platform-independent interface is that it allows greater portability between platforms. Telephony-specific software that is not otherwise using OS-specific functionality (graphical user interface support, native file management, etc.) can be ported between platforms with little or no effort. Realistically speaking, however, most applications do take advantage of OS-specific capabilities regardless of what CTI interface they happen to use, so portability benefits are limited only to the CTI portion of an application.

8.4.1. Windows Telephony: TAPI

At the center of the Windows Telephony specification (part of the Windows Open Services Architecture, or WOSA) is *TAPI*: the *Telephony Application Programming Interface*.[8-9] Microsoft includes implementations of TAPI with its various releases and versions of the Windows operating system.

TAPI allows telephony-specific applications[8-10] to access CTI functionality at a low level. TAPI defines functions for:

- Initializing and shutting down instances of the TAPI implementation. This includes the ability to register message handlers that are used by TAPI to deliver event messages to applications;

- Negotiating the appropriate version of each portion of the API that will be used by an application;

- Identifying, opening, and closing a connection to a particular switching domain (referred to as a *line device* in TAPI);

- Performing capabilities exchange, or *Caps()* functions;

- Allowing snapshots of current call or device information to be obtained;

- Establishing monitoring and specifying which events are to be filtered (and also keeping track of what monitoring has been requested on the application's behalf);

- Issuing requests for basic and supplementary telephony features and services;

8-9. TAPI documentation — TAPI is documented in the book *Telephony Application Programming Interface Programmer's Reference*, available from Microsoft through the Microsoft Developer Network Library.

8-10. TAPI support for telephony-aware applications — In Windows Telephony, telephony-aware applications are supported by a special portion of TAPI known as *Assisted Telephony*. This part of TAPI is a high-level CTI procedural interface, and is described in section 8.8.

- Detecting and generating telephony tones and DTMF tones;

- Observing and controlling the components of a physical device element (referred to as a *terminal* or *phone device* in TAPI); and

- Supporting vendor specific functionality or extended services, including versioning and exchanging private information.

Additional functionality provided by TAPI addresses issues related to running multiple applications under Windows, making use of Windows media access interfaces, and performing common application tasks. These include functions for:

- Performing housekeeping tasks such as allocating/deallocating data structures for tracking calls, and assigning particular objects to numeric identifiers.

- Allowing the CTI client implementation (telephony service providers) associated with a particular switching domain (line device) to display a configuration dialog box. This allows CTI client implementations that are not autoconfiguring the ability to be administered from within a TAPI application.

- Performing dial plan management and address translation.

- Supporting media handling and the ownership of a call by a particular TAPI client application. This allows incoming calls to be directed to specific applications that can determine the type of media stream they are carrying and perform a *handoff* to another application if appropriate. TAPI keeps track of the media type that is identified, and has extensive functionality for identifying the correct application to work with the call and the appropriate media service API to use.

TAPI's function calls and data structures are designed for use with switching domains that contain only a single logical device or device configuration, that is, ones that support only first-party call control. This significant simplification (relative to the full abstraction of telephony) allows for application implementations to be streamlined somewhat. When observing and controlling a first-party switching domain, calls, call appearances, and connections are all effectively interchangeable. With this insight, TAPI defines an *address* as the logical device in the switching domain (TAPI line device) that can be observed and controlled. A device configuration consisting of multiple logical devices is therefore represented by multiple addresses.

Applications are given visibility only to the calls associated with an address, but not to the associated appearances and connections. While this generally simplifies the programming model for most simple applications, it requires that all of the connection states associated with a particular call be reflected as a single *call state*. As an aid to programmers, call state definitions also incorporate context information. Call states can be translated to connection states as shown in Table 8-1.

Table 8-1. TAPI call states

TAPI Call State	Local Connection State	Connection state of device outside switching domain
Accepted	*Alerting* (*ringing* mode)	N/A
Busy	*Connected*	*Fail* (w/ busy cause)
Conferenced	*Connected* (after *conferenced*)	N/A
Connected	*Connected*	*Connected*
Dialing	*Initiated* (after *digits dialed*)	N/A
Dialtone	*Initiated*	N/A
Disconnected	*Connected*	*Null* (after *connection cleared*)
Idle	*Null*	N/A
Offering	*Alerting* (*offered* mode)	N/A
Onhold	*Hold*	N/A
On hold pending conference	*Hold* (consult purpose of *conference*)	N/A
On hold pending transfer	*Hold* (consult purpose of *transfer*)	N/A
Proceeding	*Connected*	Unknown [*connected*] (after *network reached*)
Ringback	*Connected*	*Alerting* (*ringing* mode)
Special info	*Connected*	*Fail* (w/other cause)

8.4.2. Macintosh Telephony Architecture: Telephone Manager

The *Telephone Manager* is the procedural interface designated by the Macintosh Telephony Architecture for writing screen-based telephony applications and most programmed telephony applications. The Telephone Manager[8-11] is distributed as part of Mac OS.

The Telephone Manager provides CTI applications with functions for:

- Identifying the appropriate version of the Telephone Manager API to use.

- Identifying, initializing, opening, and closing a connection with a particular switching domain (known as a *terminal* in the Telephone Manager API).

- Establishing, clearing, setting filters for, and tracking the status of message handlers. The Telephone Manager interface delivers messages to separate handlers according to what they reference. There are message handlers for messages applying to the whole switching domain (terminal), for logical devices (referred to as a *directory number* in the Telephone Manager), and for call appearances.

- Obtaining the capabilities of a switching domain (terminal), a physical device (associated with the terminal), or a logical device (directory number).

- Obtaining a snapshot of all the calls associated with a device or all the information about a call. In both cases this information includes dynamic feature presentation data, so the application knows what service requests apply to the device.

- Issuing requests for telephony features and services.

8-11. Telephone Manager specifications — Documentation for developing applications using the Telephone Manager is found in the Telephone Manager reference documentation that is part of the Mac OS Software Developer's Kit. This product is available from Apple through APDA.

- Detecting and generating telephony tones and DTMF tones.

- Observing and controlling the components of a physical device element.

- Accessing vendor specific functionality (referred to as *other features* and *other functions*).

The Telephone Manager also provides applications with a number of value-added functions that simplify application development. They include:

- Functions for initializing the Telephone Manager.

- A suite of functions that allow the CTI client implementation (telephone tool) associated with a particular switching domain (terminal) to display a configuration dialog box. This gives non-autoconfiguring CTI client implementations the ability to be administered from within a Telephone Manager application. A CTI Plug & Play client implementation uses this mechanism to determine what session/transport protocol stack implementation and address to use to connect to the desired CTI logical server. In addition, through the use of additional functions CTI client implementations are able to display windows and menus that coexist with the application's.

- Functions that perform housekeeping functions such as locating, allocating, and deallocating data structures for tracking calls and devices.

The Telephone Manager allows telephony-specific applications to access CTI functionality through a first-party call control model. This means that the switching domains (terminals) that applications may observe and control through this API are represented by a single device configuration, where each logical device in the device configuration is referred to as a directory number. The first-party nature of this interface also allows for a simplification in which call appearances, connections, and calls are treated as equivalent.

The relationship between the Telephone Manager's call appearance states and connection states is presented in Table 8-2. These call appearance states reflect the seven different states that the local connection can take and, for the ease of programmers, provide variations on the *alerting*, *connected*, *hold*, and *initiated* connection states to incorporate recent event information into the state itself. In addition, there are five call appearance states that correspond to a local connection state of *connected* in combination with the connection state and context of a called device

Table 8-2. Telephone Manager call states

Call Appearance State	Local Connection State	Connection state of called device
Active	*Connected*	*Connected* or *Hold*
Alerting	*Alerting* (*ringing* mode)	N/A
Busy	*Connected*	*Fail* (w/ busy cause)
Conferenced	*Connected* (after *conferenced*)	N/A
Conferenced held	*Hold* (consult purpose of *conference*)	N/A
Dialing	*Initiated* (after *digits dialed*)	N/A
Dialtone	*Initiated*	N/A
Held	*Hold*	N/A
Idle	*Null*	N/A
In use	*Fail* (w/blocking cause)	N/A
Offering	*Alerting* (*offered* mode)	N/A
Queued	*Queued*	N/A
Reorder	*Connected*	*Fail* (w/reorder cause)
Ringing	*Connected*	*Alerting* (*ringing* mode)
Waiting	*Connected*	Unknown [*connected*] (*network reached*)

8.4.3. TSAPI

TSAPI is the name of the procedural interface that is based on the standard CTI protocols. The latest version of the TSAPI specification is defined as part of the *Versit CTI Encyclopedia* (Versit, 1996) and is referred to as Versit TSAPI. Other TSAPI versions that predate Versit TSAPI are effectively subsets of it that do not support CTI Plug & Play. This book concentrates on the latest TSAPI specification, which is in the public domain and can be implemented by any vendor.

The functions making up the TSAPI interface are quite simple and easy to use because TSAPI is defined to operate on any operating system platform, and because its focus is exclusively on providing a consistent API for access to the standard CTI protocols. TSAPI does not offer extra value-added services that might mask the accurate flow of CTI control and status information. The application, or other client software, is itself responsible for interpreting capabilities exchange information, events regarding the merging of calls, and generally for keeping track of the information that is reported in the messages provided by the switching domain.

TSAPI is divided into two functional parts: ACS and CTI.

Application Control Service (ACS) functionality

ACS functionality provides applications with the ability to establish and manage a CTI stream to a desired source of CTI functionality (a logical CTI server). It includes functions to identify the appropriate API version and to locate switching domains that may be accessed. The balance of the ACS functions obtain the information necessary to communicate with a switching domain, open and close a CTI stream, and retrieve messages sent across the CTI stream.

CTI functionality

CTI functionality corresponds to the service requests, acknowledgments, and events defined for the standard CTI protocols. TSAPI provides access to CTI messages through TSAPI function calls and TSAPI application events.

TSAPI function calls correspond to messages that are to be sent across the CTI stream. Each TSAPI function is defined as generating either a specific service request or a specific acknowledgment to a service request sent by the switching domain.

TSAPI application events correspond to messages received across the CTI stream. Each TSAPI application event is defined as corresponding to a CTI event message, a positive acknowledgment, a negative acknowledgment, or a service request sent by the switching domain.

 TSAPI application events must not be confused with the CTI event messages that are sent by the switching domain. CTI event messages are one type of CTI message, whereas TSAPI application events represent any type of message sent from the switching domain.

8.5. Media Access Interfaces

The low-level programmatic CTI interfaces in the CTI software framework are complemented by all applicable media service interfaces available for use with telephony on a given platform. Regardless of the CTI interface used, applications that want to access media information associated with a particular call must use the CTI interface to identify what media service instances are available for use with a given call, and then use them to bind the desired media service to the call. Once this is done, the application uses the appropriate media service interface for the operating system being used to access the desired media on the call. The application uses a media stream identifier (or equivalent) provided by the CTI interface to specify to the media access interface what media stream is desired.

Media access is implemented in CTI Plug & Play environments using well-defined media services type values. Each media services type value corresponds to a:

- Specification for how the media stream identifier that is delivered in the CTI stream is to be interpreted in order to establish a media service stream for access to the desired media service instance.

- Specific protocol (or interpretation of data) that is to be used with the resulting media service stream.

The media service types listed in Table 8-3 are part of the CTI standard protocols. They allow basic media services to be accessed by a logical CTI client in a CTI Plug & Play fashion, that is, without requiring any implementation-specific media access drivers.

Table 8-3. Standard media service types

Media Service	Media Stream Identifier Represents
Live Sound Capture - Analog	Analog jack used
Live Sound Transmit - Analog	Analog jack used
Live Sound Capture - FireWire	FireWire channel used
Live Sound Transmit - FireWire	FireWire channel used
Live Sound Capture and Transmit - GeoPort	GeoPort stream used
Live Sound Capture and Transmit - ATM	ATM virtual channel used
Live Sound Capture and Transmit - ISDN	ISDN bearer channel used
Sound Capture and Transmit - Rockwell ADPCM	Address to be used on same data connection used for CTI Stream
Sound Capture and Transmit - API	Identifies mapper-supplied sound drivers to be used with OS API
ECTF S.100 Media Services	Address to be used on same data connection used for CTI Stream
Data/Fax Modem	Address to be used on same data connection used for CTI Stream
Digital Data - Isochronous - FireWire	FireWire channel used
Digital Data - Isochronous - GeoPort	GeoPort stream used
Digital Data - Isochronous - ATM	ATM virtual channel used
Digital Data - Isochronous - ISDN	ISDN bearer channel used
Digital Data - API	Identifies mapper-supplied drivers to be used with OS API

In CTI solutions involving mappers or API-specific adapters, these software components include (or can be tied to) specific media access drivers that are invoked when media binding takes place.

8.5.1. Mac OS Media Access Interfaces

The Macintosh Telephony Architecture specifies a number of media services and corresponding media service interfaces to be used in conjunction with telephony. These are listed in Table 8-4.

Table 8-4. Mac OS media service interfaces

Media service	Media service interface
Stream-based data	Open Transport
Fax/data modem	Connection Manager
Sound Capture	Sound Input Manager
Sound Playback	Sound Manager
Text to Speech	PlainTalk TTS
Multi-track media	QuickTime

8.5.2. Windows Media Access Interfaces

Windows Telephony specifies a number of device classes and corresponding media service interfaces that can be used to access the media streams associated with active calls. See Table 8-5 for the list of these interfaces.

Table 8-5. Windows telephony device classes and media service interfaces

Device Class	Media service interface
comm	File and communications functions
comm/datamodem	File and communications functions
wave/in	Wave functions
wave/out	Wave functions
midi/in	MIDI functions
midi/out	MIDI functions
ndis	Network driver media access control functions

8.6. Screen-Based Telephone Applications

Screen-based telephone applications (*SBTs* for short) are responsible for providing the on-screen user interface that allows an individual with a computer to manage telephone calls. SBTs represent the most visible category of telephony-specific application.

Individuals use their screen-based telephone both to do the things that they might previously have done with their physical telephone, and also to take advantage of telephony features and services that previously were never easy or possible to use. In order for the transition to using an SBT to be successful, however, the SBT must do everything that the telephone set can do and be:

- Easier to use,

- Faster to use,

- Just as reliable, if not more.

Screen-based telephone applications that meet these basic requirements quickly become the most important and most relied-upon application on people's computers. It generally is the one application that is set to launch whenever the computer starts up, and is the last application to be shut down if the computer is turned off.

The basic features of a screen-based telephone application include:

- Click-and-drag dialing

 Click-and-drag dialing involves selecting a name or phone number on the screen and dragging it to the screen-based telephone (or equivalently selecting a special dial menu item or function key) to call that number or person. Typically the SBT will support recognition and correct interpretation of number fragments. Dialing names involves having the SBT look up the numbers for people (or simply pass the names to the switching domain if it supports dial-by-name).

- Custom speed-dialing lists

    Most telephone users have a small number of people whom they call with great frequency. While many telephone sets have the ability to program speed-dial (or rep-dial) buttons, few people take advantage of this feature because of the user interface barrier. SBTs make setting up speed-dial lists easy. A good speed-dial list implementation not only lets you specify the people you want in your speed-dial list, but optionally keeps track of additional *most frequently called* and *most recently called* lists. Another feature to look for in a speed-dial implementation is the ability to keep track of all the numbers for a given person; if the person isn't in one location, you can try another without resorting to a different directory to look up the alternate number.

- Presenting callerID information

    A good screen-based telephone application will present callerID information in some fashion if it is available. It also should present the name of the person normally associated with the number, if known, rather than just the number itself.

- Customized call announcement

    A popular SBT feature is the ability of the application to provide an alternative notification mechanism for incoming calls. The SBT does this by setting the ringer volume to zero on the physical element in use (if any) using the *set ringer status* service, and then playing a customized sound or speaking a customized announcement to indicate the phone is ringing. For example, Bob's screen-based telephone application could be set to say, "Bob, there's a call for you!" every time the phone rings. By using a text-to-speech engine, the SBT can speak the name of the caller out loud using callerID information.

- Support for Set Microphone Mute

    The ability to support the *set microphone mute* service is a critical feature for a screen-based telephone. In contrast to the user of a traditional handset, who can place a hand over the microphone for privacy, the SBT user relies on the application's user interface to get immediate privacy if interrupted. If the switching domain being used does not support the *set microphone mute* service, the *hold call* service is a good fallback.

- Support for Transfer Call

  Of all of the common telephony features, transfer is the one with which people most frequently have problems. Effortless transferring of calls is something from which everyone in a business environment benefits.

Screen-based telephone applications have three basic responsibilities in the context of a CTI solution. They provide:

1. A telephony user interface to an individual

2. Full access to desired telephony functionality

3. Support for telephony-aware applications

Screen-based telephone applications should, in general, be as focused and modular as possible. Too many bells and whistles detract from the usability and universality of a product. If needed by a particular individual, these features can be added as customization using telephony-aware applications (described in section 8.8). Keeping an SBT design "focused" means that individuals will have the ability to use the screen-based telephone application that best meets their needs, without having to trade off against features that might be implemented best in a separate, complementary application. Depending on the customer, however, it might be quite appropriate to have many features in a single product. For example, a screen-based telephone to be used in a home environment might integrate screen-based telephony with voice mail (which is a programmed telephony function).

8.6.1. User Interface

User interface is the most important part of a screen-based telephone application. A computer user's screen-based telephone application in effect *becomes* their telephone, and that means it is the focal point of communications activities. The user interface design for SBTs is the place where much of the value is added.

It is very important to note that there is no such thing as the "optimal user interface design" for screen-based telephone applications. User interface studies have shown that an SBT product with a given feature set cannot have its user interface optimized for the general population (as is the case for many categories of software); this is because the expectation, work patterns, mental models for telephony, and user preferences are simply too diverse. A trip to the telephone department of any consumer electronics store underscores the fact that people have very diverse preferences when it comes to the look and feel of their telephones. This insight has a number of implications for people at different layers in the CTI value chain.

- For CTI application developers it means that there is great opportunity in developing screen-based telephone applications, because it is unlikely that any single design will dominate this field.

- Mainstream application developers adding telephony support to their applications should not assume that their product will be used in conjunction with any particular screen-based telephone application.

- Telephone network providers, telephone equipment vendors, hardware vendors, and operating system vendors considering bundling or promoting a screen-based telephone application with their products should consider structuring such deals as sampler promotions, where customers get a chance to try out a number of different SBTs. Choosing just a single SBT to work with means that a significant portion of customers will at best be ambivalent, and at worst will actually be turned off.

- System integrators and organizations implementing CTI solutions should assume that people using the CTI system will want the ability to choose their own screen-based telephone application. Solutions should be built such that individuals are given a range of SBT options from which to choose.

- Individuals should be choosy. If the screen-based telephone application you're using doesn't feel right, then try another one.

Another important insight into screen-based telephone applications is that users have (and need) only one. Though this might seem to be an obvious statement, some application designers overlook it. Just as people rarely have multiple telephone sets on their physical desks, they need or can effectively use, only one SBT on their virtual, computer desktops. The implication for application developers reinforces the fact that SBT functionality should be narrowly focused. Some mainstream developers, in trying to add telephony support to their applications, actually turned them into screen-based telephone applications rather than telephony-aware applications. This approach really isn't practical because the user would end up with multiple SBTs if other mainstream applications started doing this.

When considering options for screen-based telephones (or when designing them), keep in mind that an SBT needs to be lightweight in its use of memory and other computing resources. In order for users to have a positive experience, they need to be able to rely on their application being available at all times; that means it has to be in memory all the time.

In this section we'll look at a number of different user interface paradigms for screen-based telephone application design. The SBTs that you evaluate (or build) may fall cleanly into these categories; more likely, however, they will contain elements from multiple schools of design. Again, it is important to keep in mind that there is no right answer when it comes to SBT design. There is only the need for sufficient diversity to satisfy the preferences of a diverse population.

Figure 8-13. Phone-under-glass design approach

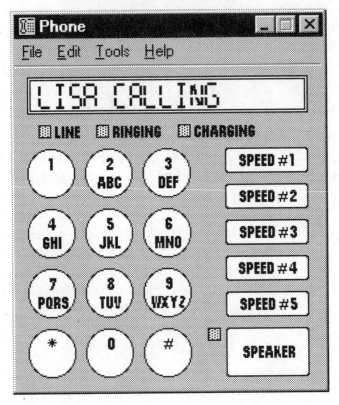

Phone Under Glass

One approach taken in the design of SBTs is to make the on-screen view of the virtual telephone look and feel much like the physical telephone that it replaces. This is referred to as putting the *phone under glass* (Figure 8-13).

The thinking behind this approach is that users get all the benefits of screen-based telephony described above, but they still continue to work with an interface with which they are familiar. The traditional rough edges of a phone's design can be polished through online help dialogs and help balloons that explain what all the buttons do.

Designers who don't like this approach point out that it preserves the limited user interfaces that aren't very effective anyway, and it results in the consumption of much more screen real estate than is really necessary. Another criticism is that users don't want to click on-screen buttons representing a dial pad when they can just type numbers very quickly on the keyboard. As with all of these schools of thought, there is no single right answer.

Figure 8-14. Button panel design approach

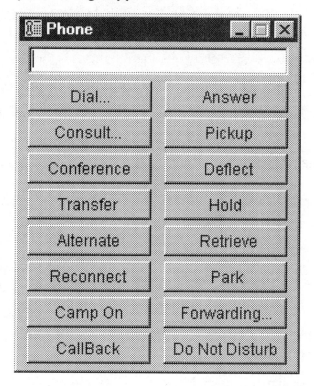

Button Panel

The *button panel* approach to SBT design involves putting all of the features and services that are accessible using the SBT in a window or menu of some kind as an array of buttons, along with a simple text area that provides status information. When a user wants to place a call, for example, he or she just presses the "Dial..." button and is prompted for the number to dial. A number then can be typed or dragged from somewhere (see Figure 8-14).

In this type of design, the window or menu may be hidden from view (collapsed or actually absent from the screen) until needed. Upon the arrival of an incoming call, it might appear automatically with the "Answer" button highlighted as the default and the display area providing the callerID. For outgoing calls, a function key or "hot area" for the mouse on the computer display could be used to pop the SBT to the foreground.

Proponents of this design style maintain that this is the best approach because every option is plainly labeled and it is very efficient in terms of screen real estate. Dissenters say that it is good only for linear thinkers because of its reliance on text labels and a text status display. Again, there are no right answers.

Figure 8-15. Minimalist design approach

Minimalist

Yet another design approach is the graphical minimalist approach. In this approach, the emphasis is on minimizing screen real estate but using a graphical rather than text view. A postage-stamp-sized window contains an icon that indicates the status of any telephony activity. The window floats in front of all other windows so that it is always accessible, but it is tiny so it doesn't get in the way. If there is an active call, the icon reflects that status and the name(s) and phone number(s) of the participants appear. Calls can be placed (or added to) at any time by dragging telephone numbers or names to the window. Services and features can be invoked using simple mouse gestures and pop-up menus. This design approach is illustrated in Figure 8-15.

Supporters of this approach say that it strikes the best balance of screen real estate and easy-to-understand graphics. Detractors find fault in the fact that the capabilities of the application are not readily apparent because the menu must be popped up and the gestures must be learned.

Figure 8-16. Direct manipulation/visualization design approach

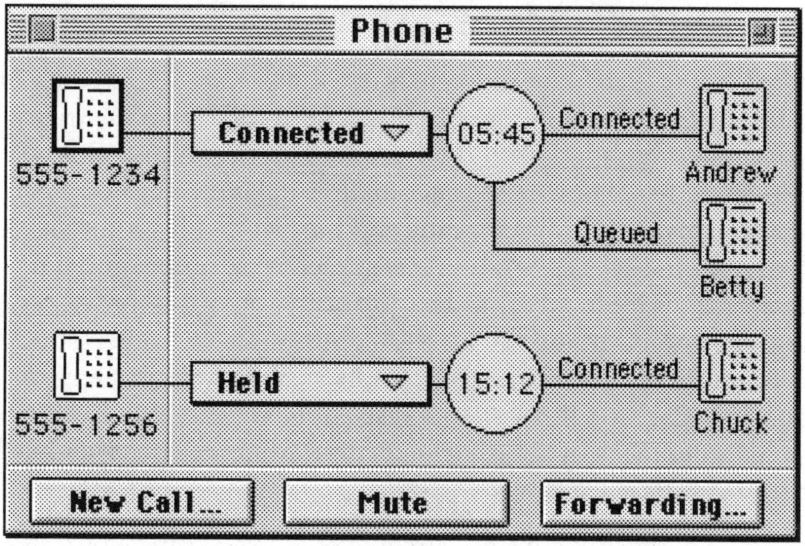

Direct Manipulation or Visualization

The direct manipulation/visualization approach involves changing the user paradigm for telephony from the traditional, feature-oriented view to a view that reflects what is really taking place in the switch. The user interface is based on a graphical representation of the devices, with calls being used much as they are presented in Chapter 4. The users of this type of SBT are able to see the devices they control, the connected calls, and the other devices in these calls. They can use the mouse to manipulate connections directly, without having to think in terms of actual feature or service names. For example, to add a person to a two-party call, the user just drags the name of the person to the icon representing the call (rather than pressing a button to invoke the *single step conference call* service). Figure 8-16 shows an example of this design approach.

When this approach was user-tested, a surprisingly large group was very enthusiastic. An equally large group hated it utterly. As with all of the design approaches, each is good for a certain part of the population.

Speech Interface

The last example we'll look at here is one that switches from the visual interface to the audio interface. In this case the SBT's primary user interface is through speech and other audio input/output. This type of screen-based telephone has a very small visual indicator in the corner of the screen, just to indicate status and the fact that it is running. The primary interface for controlling it is speech recognition, and the primary means for feedback is text-to-speech and prerecorded sound clips. When an incoming call arrives, for example, the caller's name is announced. To answer it, the user simply says, "Computer, answer the phone." To place a call, the user just says, "Computer, call Frederic." The computer looks up Frederic in its preconfigured speed-dial list and initiates the call.

As with all of the other examples, some people like this approach and others dislike it intensely.

8.6.2. Functionality and Feedback

A second area of responsibility for screen-based telephone applications is functionality. The range of functionality supported by different SBTs will vary, depending upon the type of user their designers had in mind. It is important to note that not every user wants access to the full range of telephony features. Nonetheless, a good screen-based telephone application doesn't make arbitrary decisions about what subset of telephony features and services it chooses to support. A screen-based telephone designed for mainstream use should support the full range of telephony features and services, as well as supporting control of physical device elements. When the application connects to the switching domain, it should then scale back its user interface to reflect the actual features and services available. A design that provides for users who prefer a reduced feature set might additionally allow users to deactivate other features and further scale back the functionality that appears in the user interface.

As we have already seen, observing is at least as important as controlling. In the context of screen-based telephone applications, this translates into putting at least as much emphasis into the implementation of feedback mechanisms as into control. Regardless of the user interface design employed, an SBT must squeeze out every ounce of information about the status of relevant items in the switching domain, and it must provide this information as feedback in an immediate fashion.

8.6.3. Support for Telephony-Aware Applications

The third area of responsibility for screen-based telephone applications is supporting telephony-aware applications. From the perspective of the CTI value chain, the screen-based telephone application is near the top. From the perspective of an individual user, however, the screen-based telephone application is the foundation and everything below it is transparent. Individuals see the screen-based telephone application as the starting point for the customized, integrated telephony workspace they will form using telephony-aware applications and scripts. A good screen-based telephone application is designed to support this role as an anchor point in a larger CTI solution. (Telephony-aware applications and the high-level programmatic CTI interfaces that support them are described in section 8.8.)

8.7. Programmed Telephony Applications

Programmed telephony applications are the complement to screen-based telephone applications. Where screen-based telephone applications provide an interface to the user who is placing calls using a particular computer, programmed telephony applications present a telephone-based interface to someone who is calling a particular computer. Where screen-based telephone applications are designed to operate under the direct control of a computer's user, programmed telephony applications are intended to run in an autonomous fashion once they have been given their instructions (or have been *programmed*).

Programmed telephony applications are telephony-specific applications that interact with telephone calls in an autonomous fashion based on predetermined rules of any complexity. A programmed telephony application could be as simple as a program that answers calls after 10 rings and tells the caller to try again at a later time. On the other hand, a more complex programmed telephony application might be a call routing product that interacts with callers using speech recognition technology and, after asking a series of questions and interpreting the replies, directs each call to the person most likely to be able to help a given caller. In fact, the possibilities for programmed telephony are limited only by the designer's imagination and the sophistication of available technology.

In carrying out their tasks, programmed telephony applications capture and use one or more of the following pieces of information relative to a call:

- Call associated information (e.g., ANI, callerID, DNIS, correlator information, and user data)

- Date, day, and time of day

- Resource availability (e.g., whether an agent is ready or not)

- DTMF tone or dial pulse information

- Recorded audio

- Speech recognition

Information captured from the call, along with information from databases and the rules that govern the operation of the programmed telephony application, are combined to determine what actions the application is to take.

The first three items on the list above do not require any interaction with the caller. The latter three involve prompting the caller in some fashion to obtain the desired response. *Voice processing* applications are the subset of programmed telephony applications that involve this type of interaction with callers. Information and prompts provided to a caller might use:

- Prerecorded (digitized) sound

- Text-to-speech

- Concatenated speech

Information also can be returned to callers using methods that include fax, either within the call itself or on a separate call; data transfer using simultaneous voice and data technology (assuming the caller has the appropriate software running); or electronic mail.

8.7.1. Programmed Telephony Application Categories

Programmed telephony applications can be categorized generally as *call logging* applications, *call routing* applications, *messaging* applications, *information access/capture* applications, and *notification* applications.

Call Logging

The simplest form of unattended programmed telephony application is one that observes only. A *call logging* application does not interact with the call except to extract the information that it will record.

Call Routing

Call routing applications comprise a class of programmed telephony solutions that automate the routing of calls. Though these applications typically take advantage of the routing services provided by the switching domain (described in Chapter 6, section 6.10). However they also might use the *single step transfer call* service, *redirect call* service, or any other call-control services.

Calls can be routed purely on the basis of call associated information, or through interaction with the caller. For example, a call routing application might route all calls from a particular area code to one group of agents during the day and to another group during the evening. In another example, this time involving interaction, a customer calls a technical support line and the programmed telephony application asks for the customer's ID number. If the ID number has expired, the customer's call is directed to the sales department. If the number is good, the call is directed to the technical support group.

Other examples of call routing applications include:

- Auto-attendants
- Call screening
- Selective blocking
- ACDs

Messaging

Messaging is the most ubiquitous example of a programmed telephony application. Messaging involves:

- Interacting with the caller to identify the destination of the message, if necessary. (In most cases this is not necessary because it is either implicit or is provided through call associated information.)
- Playing an appropriate message prompt or greeting.
- Capturing the voice information representing the message.

- Forwarding the message to the appropriate mailbox.

Messaging also may apply to the receipt of fax documents.

Information Capture

Information capture applications are those that are designed to capture from callers information beyond a simple message. Examples of information capture systems include:

- Order entry (i.e., a caller is buying something)

- Dial-in questionnaires and surveys

Information Access

Information access applications are programmed telephony applications that allow callers to specify and retrieve information interactively. Examples include:

- Fax-back applications

- Database retrieval

- Document retrieval

- Message retrieval

Notification

Notification applications are programmed telephony applications designed to initiate calls at predetermined times or under certain conditions. For example, a volunteer fire department might set up a notification application to handle notifying all of the volunteer firefighters. After being triggered with the information about a fire, this application calls each volunteer and plays back the message describing the fire and its location. Other examples include:

- Wake-up call system

- Remote network or system monitoring

- Reminder of overdue items

8.7.2. Commercial Programmed Telephony Applications

Commercially available programmed telephony applications fall into three basic categories: single-purpose products, customizable products, and application generators.

Single-purpose Products

Single-purpose programmed telephony applications are those whose programming, or the logic governing their operation, are fixed and not subject to further customization by the computer user. These systems generally do provide some means for specifying particular parameters, however.

For example, an answering machine application generally is a single-purpose programmed telephony application. The product allows customization of parameters such as the greeting that is played, the number of rings to wait before answering, and the maximum length of a message. All of the logic that drives the answering machine application is not subject to customization, however; it is hard-coded by the developer of the application.

Customizable Products

Customizable programmed telephony applications are those that are intended for a specific purpose, but allow the customer to define rules that govern its operation, based on a particular set of options.

For example, an auto-attendant application might provide the complete framework and all of the normal algorithms used by an auto-attendant, things such as prompting for a number, looking up an address, transferring calls, parking calls, and playing messages. Final assembly of the logic, however, is left up to the customer. The customer not only records all of the messages (the parameters for the system), but also decides on the logic or rules that will be used for navigating through the automated attendant system.

CTI Application Generators

A *CTI application generator* is a programmed telephony application that provides a complete range of fundamental programmed telephony building blocks, but the application ships unprogrammed. The product is designed to allow the customer to assemble the software building blocks in any arbitrary way for a given solution. The customer is essentially purchasing an erector set that is then used to assemble the desired solution.

CTI application generators are really software development tools in their own right, and they come in many different forms. The easiest to use are those based on graphical programming languages and use icons and connecting lines to show the logical flow between actions the application will take.

CTI application generators are among the most powerful of the commercially available CTI products because they are effectively anything the user wants them to be. With a little bit of effort, one can build any programmed telephony application using these software components.

8.7.3. User Interface Considerations

User interface considerations are extremely important in the design of a programmed telephony application. In this context, however, the user with whom we are principally concerned is the person at the other end of the call, not a computer user. The person who is interacting with a programmed telephony application has a very limited interface:

- The person hears sound generated by the programmed telephony application.

- The person can dial digits.

- The person can speak.

As a result, designing the other-end interface presented by a programmed telephony application can be much more challenging than designing a visual interface for the CTI user, where it is much easier to provide feedback and many more options can be presented simultaneously. This section presents a number of things to keep in mind when evaluating or building a programmed telephony application.

Consistency

The most important consideration in user interface design, regardless of the context, is *consistency*. A programmed telephony application should behave in a consistent fashion no matter where it might be in its logic flow. This provides callers with a sense of security and reliability that makes them much more comfortable about interacting with an automated system. Examples of consistency rules include:

- Reaching an operator

  A single digit should be reserved for reaching a live operator (if appropriate for the application in question). There should be just one value (typically it is "0" or "*") used throughout the application. If the caller presses the designated digit at any time (except perhaps in the middle of entering a multi-digit number), the call should be transferred to a designated operator. If no operator is available, an appropriate message should be played. (For example, after hours on weeknights the message might say, "There is no one available to take your call at the moment. Please leave a message and we'll return your call tomorrow.")

- Reaching the main menu

  Most programmed telephony applications present a hierarchy of menus and allow callers to navigate by pressing digits that identify a desired choice. The root of this hierarchy, the starting point, is referred to as the *main menu*. A single digit may be reserved for returning to this main menu at any time. If this feature is supported, the same key must be used throughout the program and the option must be available at all times.

- Number entry format

    Programmed telephony applications often prompt callers to enter a number of some sort, such as a credit card number, account number, product number, etc. The application should always use the same methodology for collecting a number from the caller. The best approach generally is to use numbers that have fixed numbers of digits. The prompts in these cases would say something along the lines of, "Please enter the six-digit account number now." The caller would then enter six digits (or a time-out would occur).

    Another methodology allows entry of a variable number of digits by using the "#" key to indicate the end of the sequence. (This has become a de facto standard as a result of its use in paging systems.) In this case the prompt would say something along the lines of, "Please enter the account number followed by the pound sign."

    All spoken prompts for a user to enter a number should be identical so as to reinforce that the same methodology is being used consistently. The application might supplement this prompt with a tone (such as the bong tone).

- Date format

    Another type of data that callers are frequently asked to provide is a date. Dates always should be entered numerically (as opposed to using the letters printed on the dial pad buttons). Whatever sequence of prompts is used should be consistent throughout the application.

- Canceling an entry

    The application should expect that callers frequently will make mistakes when using this primitive interface. While supporting a backspace mechanism is unrealistic, a single digit should be used throughout the system to cancel an entry and start over. A good digit for this is the "*". When the designated digit is pressed, the system should discard the partial number or date and play a variation of the last prompt that incorporates the word "re-enter."

- Fail-safe confirmation

  Any time that a significant entry is to be made that cannot be undone, there should be a confirmation step to ensure that the last request was not entered by accident. The consistent support for this confirmation step represents a safety net that will make users much more confident in how they use the application.

Always Allow Interruption

Once a given caller has become familiar with the operation of a particular program he or she will want to move more quickly, without listening to each prompt in its entirety. It is therefore important that programmed telephony applications be prepared to interrupt the playback of a prompt if the response has already been received.

Support for *type-ahead* means that a user can enter a sequence of responses in anticipation of a prompt, so the prompt may be skipped entirely. Type-ahead should be supported except in the case of confirmations.

Presenting Spoken Information

When presenting information in spoken form, it is very important that the most important, desired, or unique information be spoken as early as possible in each statement. The key is to allow the caller to comprehend the desired information as quickly as possible, without abbreviating the language used. Statements should be complete sentences in order to sound natural and to provide the redundancy that is important in spoken communication.

An example illustrating the application of this principle might be the choice of statements in a system that provides order status. If the enquiry was to determine if a particular order was shipped, a system might say, "Your order, #12345, consisting of 2 items was shipped today at 12:02." This could be improved by saying instead, "At 12:02 today your order, #12345, consisting of 2 items, was shipped." Given the likelihood that the person already knows the order number and the number of items ordered, the fact that it shipped should go first. In this example it is also a good idea to have a very different-sounding phrase to state that the order was not shipped. For example, the statement "Order #12345 has not yet been shipped" allows the listener to identify the difference between the two possible cases after hearing just the first syllables of the response.

This principle is applied to menus by listing items in the order of frequency of use; callers can press the digit for the option desired as soon as they hear it, minimizing the time spent listening to prompts. On a related note, the digit corresponding to a particular menu choice should follow the item, not precede it. This frees the listener from having to remember each number while listening to each description.

Assume a Noisy Line

The design of a programmed telephony application always should assume a noisy phone line and therefore the possibility that a caller might be having difficulty making out what is being said. In particular, callers from cellular phones drop out of a call for a few seconds at a time. Especially in cases where a name of some sort (person's name, company name, street name, etc.) is being read back, it is very important to allow the caller to request that the statement be played again. In fact, it is highly desirable to allow the caller to request something be spelled out letter by letter, if necessary using disambiguators (such as the phonetic "N as in Nancy").

Assume Errors

A good programmed telephony application will devote at least half of its code to handling errors. It is certain that, at every step in the logic, something other than the correct input will be received. A good design will anticipate all of the possibilities. This makes for a robust solution and a positive user experience.

8.8. Telephony-Aware Applications

Telephony-aware applications are mainstream applications or pieces of solution software that have been enhanced to support telephony integration by using high-level programmatic CTI interfaces.

Both the Macintosh Telephony Architecture and Windows Telephony define high-level programmatic CTI interfaces. The Windows Telephony approach involves a special, high-level portion of TAPI known as the Assisted Telephony API. The heart of the Macintosh Telephony Architecture is its high-level CTI object-oriented interface, known as the Telephony Apple Event suite.

The capabilities of the two interfaces differ, so the range of functionality available to telephony-aware applications is not the same on both platforms. However, a generic example of how a telephony-aware application works on either platform is as follows:

- Andrew's computer is running a screen-based telephone application and a number of productivity applications.

- An expense form document appears in Andrew's electronic mailbox for his approval. He opens it to find that the form contains a number of questionable expenses, so he decides to call the person who completed the form.

- The expense form application he is using happens to be telephony-aware, so all Andrew has to do is select the "Call sender" menu item from the application menu bar. Using the information it already has built into the form, plus other databases to which it has access, the application sends a command to place a telephone call to the sender of the form.

- The command to place the phone call is transferred through the high-level CTI interface to the screen-based telephone application.

- The screen-based telephone application places the call as if Andrew had activated it directly, looking up the appropriate number and dialing it using the screen-based telephone application itself.

The telephony-aware application is responsible only for initiating new calls and other related telephony tasks. As always, the screen-based telephone application is responsible for managing calls and providing feedback.

Applications that support built-in scripting languages, if their scripting engine provides access, need not support the high-level CTI interface directly in order to become telephony-aware. The same scenario described above could have been accomplished by having the user trigger a script within the application.

Telephony-aware applications typically represent telephone numbers using canonical phone number format (as described in Chapter 3, section 3.10.4). Canonical phone number format is a location-independent telephone address representation that allows telephone numbers to be correctly translated by a telephony-specific application, or by the switching domain implementation at the time the call is dialed.

8.8.1. Windows Telephony: Assisted Telephony

Windows Telephony defines a set of API functions that support what is referred to as *Assisted Telephony*. For developers of mainstream applications, Assisted Telephony allows simple telephony integration support by way of a single program function (*lineRequestMakeCall*) that sends call-placement requests to a telephony-specific application designated for handling such requests. The Assisted Telephony API also includes a second function (*tapiGetLocationInfo*) for returning default location values that a telephony-aware application can use to convert noncanonical numbers into canonical phone number format.

The remaining function calls associated with the Assisted Telephony API are used by the designated telephony-specific application that is handling requests. These are used to register for receiving requests and to retrieve requests.

8.8.2. Macintosh Telephony Architecture: Telephony Apple Events

At the heart of the Macintosh Telephony Architecture is the specification for a suite of *Telephony Apple Events*. Apple Event technology is a capability of Mac OS that supports the exchange of object-oriented messages using a well-defined protocol (referred to as the *Apple Event Interprocess Messaging Protocol*). Apple Events is an object-based programmatic interface, so on the Macintosh operating system the high-level CTI interface is actually object-based.

The telephony Apple Event suite models the switching domain by means of an object, representing the switching domain, that contains one or more directory number objects. These in turn contain zero or more call appearances (at any given time). Telephony Apple Event messages travel between the telephony-aware application and the screen-based telephone application; they describe changes in, or to be made to, the attributes of these objects. The Apple Events that are defined as part of the Telephony Apple Event suite and that manipulate objects include:

- Answer Call

- Conference

- Dial Digits

- Drop Call

- Forward

- Hold Call

- Make Call

- Park Call

- Redirect

In addition, the following Telephony Apple Events are defined to establish monitoring of the objects. These notification-related events allow the telephony-aware application to specify a set of criteria which, if later satisfied by a change in the screen-based telephone application, trigger a notification event to be sent to the telephony-aware application. This mechanism can be established, for example, to log all calls that are not answered, or to notify the application of external incoming calls.

- Notify Dependency

- Register Dependency

- Release Dependency

AppleScript is the system-wide scripting language on the Mac OS. It is an English-like language that allows sequences of Apple Events to be stored as scripts. AppleScript allows any end user to build a telephony-aware applet using a few lines of AppleScript.

Basic Telephony Apple Event Support

The basic category of Mac OS based telephony-aware applications simply use the make call capability, and are equivalent in functionality to applications using Assisted Telephony in Windows. These applications use the *make call* telephony Apple Event to send a message requesting that the screen-based telephone application call the person indicated by the canonical telephone number provided.

Extended Telephony Apple Event Support

Telephony-aware applications that support additional Telephony Apple Events are able to invoke many supplementary services. A good example of an application using that rich feature set is a calendar application for keeping track of telephone conference calls. At the appointed time for a conference call, it places calls to all the participants and conferences them all together. As with all telephony-aware applications, the screen-based telephone application is responsible for actually managing the resulting call(s).

Notification Telephony Apple Event Support

Telephony-aware applications that support the notification events can perform a number of tasks that augment the work performed by a screen-based telephone application. For example, a database application that supports Telephony Apple Event notifications can register with the screen-based telephone application and await incoming external calls. When such a call is received, the screen-based telephone application informs the database application of the new call and its associated callerID information. The database then is able to use this information to automatically find and display records that pertain to the caller—even before the call has been answered.

User Scripted

Mainstream applications that do not support Telephony Apple Events internally, but that may be scripted using AppleScript, can be made telephony-aware by writing scripts that both drive the application and send Telephony Apple Events to a screen-based telephone. These scripts then can be attached to the application or to specific documents used by the application.

AppleScript Application

Finally, entire applications can be built using AppleScript. A telephony-aware application easily can be built from scratch in this fashion. A typical use of such an approach might be building the custom software for a call center agent's work environment. Each agent's computer runs a screen-based telephone, the AppleScript-based application, and other utility applications that are launched as needed. When a call is directed to the agent, the AppleScript-based application acts as the front end. When it receives the notification Apple Event for the call, it opens the necessary databases and other files the agent needs. The AppleScript application acts as a dashboard for the call center agent.

8.8.3. Proprietary High-Level Interfaces

In addition to the OS-defined high-level programmatic CTI interfaces, various screen-based telephone application developers have also developed their own proprietary mechanisms for integrating with mainstream applications.

8.9. Creating Custom CTI Solutions: Using Off-the-Shelf Software

In many ways, the truly exciting payoff from the technologies described in this book is the opportunity created for system integrators, customers, and end users to rapidly build powerful, customized CTI solutions. The technologies previously described for conveying CTI messages between components, and ultimately delivering them to various applications on computer systems, simply lays the foundation for those who actually derive the value of this infrastructure.

The maturation of standards for CTI Plug & Play connectivity and well-defined programmatic APIs will lead to larger and larger numbers of off-the-shelf telephony-specific and telephony-aware applications. The ease of creating telephony-aware applications on either of the mainstream personal computer platforms will mean an abundance of these products.

For all of those on the upper end of the CTI value chain, this means two things:

1. Extremely powerful solutions can be built using the existing generations of screen-based telephony applications, programmed telephony applications (particularly application generators), and telephony-aware applications. Using scripts and other application integration tools, it is possible to integrate a number of off-the-shelf applications into a single, powerful CTI solution.

2. It will become increasingly easy to find off-the-shelf telephony-aware applications that require little or no additional customization.

8.10. Review

In this chapter we have seen the various layers that make up the *CTI software framework*. This framework describes the relationships among the various *types of CTI software components* that exist in an open computer system capable of supporting CTI solutions.

CTI software component types include *R/W interfaces* and *session/transport protocol stacks*, *software mappers*, *CTI server implementations*, *CTI client implementations*, *low-level* and *high-level procedural* and *object programmatic CTI interfaces*, *telephony-specific applications*, and *telephony-aware applications*.

R/W interfaces used to connect to logical CTI servers provide direct control over, and access to, CTI streams. The session/transport protocol stacks used by R/W interfaces may be actual protocol stacks or software mappers that convert from proprietary protocols in a transparent fashion.

CTI client implementations are responsible for providing access to one or more switching domains through one or more programmatic CTI interfaces. *CTI Plug & Play client implementations* accomplish this by connecting to logical CTI servers using standard CTI protocols. The alternative approach is to use *API-specific adapters*, which must be provided by the vendor of a particular switching domain implementation, and which work exclusively with a low-level programmatic CTI interface on a particular operating system. With TAPI on Windows, these are referred to as *telephony service providers*. With the Telephone Manager on Mac OS, they are referred to as *telephone tools*.

Low-level programmatic CTI interfaces are designed to support the needs of CTI software developers in writing telephony-specific applications. *TSAPI* is a simple interface that maps directly to the standard CTI protocols without loss or embellishment. *TAPI* and the *Telephone Manager* are operating system vendor-provided low-level

CTI interfaces for the Windows and Mac OS operating systems, respectively. They provide simplifications for developers at the loss of some functionality and veracity.

Telephony-specific applications are either *screen-based telephone applications (SBTs)* or *programmed telephony applications*. Screen-based telephone applications are products that provide computer users an alternative user interface to the telephone set for managing telephone calls. Programmed telephony applications are products that interact with telephony resources in an autonomous fashion once they are configured or programmed with the rules that define their operation. Commercial programmed telephony applications are categorized as being *single-purpose, customizable*, or *CTI application generators*. The latter are full-fledged development tools that can be used to build any type of programmed telephony application.

High-level programmatic CTI interfaces allow telephony-aware applications to request telephony services from screen-based telephone applications. Telephony-aware applications, which by definition include any application that uses the high-level CTI interfaces, request appropriate telephony services (such as placing calls) from the screen-based telephone application. The SBT then provides the user interface for ongoing management of calls.

The benefit of this multi-layered software framework is that CTI system integrators, customers, and individuals are able to integrate off-the-shelf software products into CTI solutions. While already easy to do at present, this integration actually will become easier as more products become available.

9.
CTI Solution
Examples

We have now explored all the key concepts governing the implementation of telephone systems, the core telephony features and services, and the specific trade-offs with telephone products. We have also learned about CTI concepts, CTI configurations, and CTI software tools and interfaces. It is therefore time to pull everything together and look at how the scenarios presented in Chapter 2 are actually built. (You may wish to go back and review Chapter 2, sections 2.3 through 2.10 at this point.)

Each of the scenarios described earlier can be implemented in many different ways. Just one of the possible solutions will be presented in this chapter for each scenario, in order to demonstrate how the technologies we have explored in this book can be applied. Each solution involves a description of the corresponding CTI system configurations and the software components used. To keep the solutions as simple as possible, standard CTI protocols were used in each case. As we have seen, however, proprietary protocols can be accommodated in most situations using a variety of approaches.

9.1. Screen-Based Telephony

Andrew, the PR manager for a large public relations agency, set up a CTI solution for himself. It involved connecting his computer directly to his telephone station and using his off-the-shelf contact manager along with screen-based telephone application software. Andrew's scenario is summarized in Table 9-1. (Refer to Chapter 2, section 2.3 for the complete scenario description.)

Table 9-1. Screen-based telephony solution scenario

Name:	Andrew
Occupation:	PR manager
Location:	Corporate office
Switch:	PBX
Telephone station:	Proprietary digital telephone
Type of computer:	Desktop
Type of CTI Interface:	Direct-connect
Applications used:	Screen-based telephone
	Contact Manager
Implemented by:	Andrew
Customization:	N/A
Future Plans:	Implement on PDA for cellular phone and home phone
Benefits:	Greater professionalism and responsiveness
	Realize better utilization of computer and telephony resources
	Improves efficiency and reliability
	Improves employee morale
	Reduces staff training times
	Inexpensive: uses off-the-shelf products

9.1.1. CTI System Configuration

The CTI system configuration for Andrew's CTI solution is shown in Figure 9-1. Andrew implemented the solution by himself; he didn't have access to the switch, so a client-server solution was out of the question (and also outside his budget). Fortunately, the vendor of the switch that his company uses manufactures a telephone station peripheral for their proprietary digital lines. This is hooked up between the switch and the telephone station as shown. The telephone station peripheral connects to Andrew's computer using an RS-232 cable.

Figure 9-1. Screen-based telephony solution scenario configuration

Figure 9-2. Screen-based telephony solution scenario software

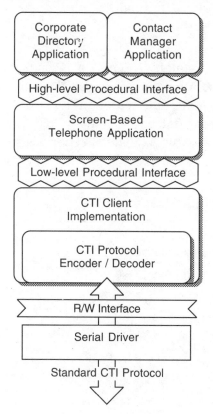

9.1.2. CTI Software Components

The software components that complete Andrew's CTI solution are shown in Figure 9-2. The telephone station peripheral supports standard CTI protocols, so no proprietary software was needed for Andrew's computer. Andrew chose the screen-based telephone application that best fit his needs and preferences; the application connected to the CTI functionality provided by the telephone station peripheral through the CTI Plug & Play client implementation installed on his system. Finally, both his contact manager and corporate directory database applications were already telephony-aware, so they were immediately able to take advantage of his new screen-based telephone application once it was installed.

9.2. Mobile CTI

Betty, the sales representative on the go, is the happy user of a sales-force automation solution that makes good use of CTI capabilities. It is centered on a sales-force automation utility application that integrates customer and product databases, along with a screen-based telephone application and fax software. Betty's scenario is summarized in Table 9-2. (Refer to Chapter 2, section 2.4 for the complete scenario description.)

Table 9-2. Mobile CTI solution scenario

Name:	Betty
Occupation:	Sales representative
Location:	On the road - hotel - airport - taxi
Switch:	Various
Telephone station:	Various
Type of computer:	Notebook
Type of CTI Interface:	Direct-connect (PC-Card modem)
Applications used:	Sales-force automation application
	Customer and product databases
	Electronic mail software

Table 9-2. Mobile CTI solution scenario (Continued)

	Fax software
	Screen-based telephone
Implemented by:	System integrator
Customization:	Preferred a different screen-based telephone
Future Plans:	Infrared connections to pay phones and SVD
Benefits:	Eliminate drudgery
	Fast, easy, and error-free dialing
	Calls placed using least expensive method
	Greater professionalism, consistency, and responsiveness
	Morale improvement

9.2.1. CTI System Configuration

The CTI system configuration for Betty's CTI solution is shown in Figure 9-3. It takes advantage of the primitive—and yet functionally sufficient—capabilities of the PCMCIA fax modem that is installed in her notebook computer. The system integrator who built the solution chose this particular modem because it supported the standard CTI protocols in addition to supporting the standard, Hayes-compatible AT commands. Using CTI Plug & Play components rather than API-specific code allows for components to be exchanged by users of different platforms, and also allows use of infrared and other devices as they become available.

Figure 9-3. Mobile CTI solution scenario configuration

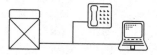

9.2.2. CTI Software Components

The software components that complete Betty's CTI solution are shown in Figure 9-4. From Betty's perspective, the principal interface to the solution is the sales-force automation utility application, written

511

Figure 9-4. Mobile CTI solution scenario software

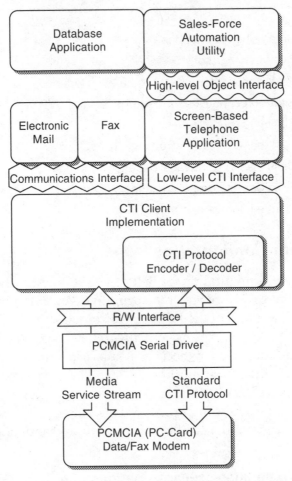

by the system integrator using high-level scripting tools. This interacts with the various database, fax, and electronic mail software that are included in the solution by means of platform-specific scripting techniques. It integrates with the screen-based telephone application using the platform's high-level programmatic CTI interface. The CTI client implementation uses standard CTI protocols to access the modem. It also has a value-added feature: It supports dial-plan management and can translate between canonical phone numbers (provided by the screen-based telephone application) and the appropriate dialing digits for the given location.

9.3. Power Dialing

Chuck, the individual working in the accounts receivable department, now uses a specially developed screen-based telephone application that includes predictive dialing to track down customers with past-due accounts. Chuck's scenario is summarized in Table 9-3. (Refer to Chapter 2, section 2.5 for the complete scenario description.)

Table 9-3. Power dialing solution scenario

Name:	Chuck
Occupation:	Accounts receivable
Location:	Corporate office
Switch:	PBX
Telephone station:	Proprietary digital telephone
Type of computer:	Multi-user
Type of CTI Interface:	Client-server
Applications used:	Accounting software
	Special predictive dialing SBT
Implemented by:	Corporate IS department
Customization:	N/A
Future Plans:	Personal computer-based system
Benefits:	Eliminate drudgery / improve morale
	Speeds work process
	Saves money and time
	Easy to develop and use

9.3.1. CTI System Configuration

The CTI system configuration for Chuck's CTI solution is shown in Figure 9-5. The accounting software currently used by Chuck's group runs on a multi-user computer as shown. It connects to a CTI server which, in turn, has access to the CTI interface on the PBX being used.

Figure 9-5. Predictive dialing solution scenario configuration

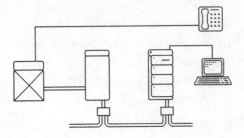

Figure 9-6. Predictive dialing solution scenario software

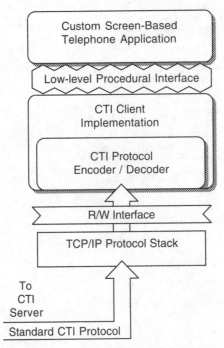

9.3.2. CTI Software Components

The software components that complete Chuck's CTI solution are shown in Figure 9-6. In this case, one of the standard CTI protocols is being delivered from the CTI server using a TCP/IP connection. The multi-user computer is not running one of the common personal computer operating systems, so there are no high-level CTI APIs or off-the-shelf CTI software. For this solution, the corporate IS department has custom-written a simple screen-based telephone application that works in conjunction with the accounting system and utilizes the low-level programmatic CTI interface exposed by the CTI client implementation. This software is built around use of the switch's *make predictive call* service.

9.4. Personal Telephone System

Debbie, the work-at-home consultant, set up a CTI solution for herself that involved installing a personal PBX for her home. By maximizing the utilization of two telephone lines, she saved the expense of having to install a third phone line. The personal PBX also provided a number of other significant benefits. Debbie's scenario is summarized in Table 9-4. (Refer to Chapter 2, section 2.6 for the complete scenario description.)

Table 9-4. Personal telephone system solution scenario

Name:	Debbie
Occupation:	Consultant
Location:	Home office
Switch:	Personal PBX
Telephone station:	Analog POTS telephone
Type of computer:	Home computer
Type of CTI Interface:	Direct-connect
Applications used:	Client database
	Time and billing database
	Voice mail software
	Call logging software
	Screen-based telephone
	Fax software
Implemented by:	Debbie
Customization:	N/A
Future Plans:	Digital system

Table 9-4. Personal telephone system solution scenario (Continued)

Benefits:	Makes home business possible
	Saves money
	Saves time
	Easy to develop and use
	Provides privacy and enhances professional image

9.4.1. CTI System Configuration

The CTI system configuration for Debbie's personal PBX solution is shown in Figure 9-7. Debbie's personal PBX connects to the two analog lines coming into her house. (She had the original line switched from a loop-start to a ground-start line and ordered the new line as a ground-start line.) All the telephones in the house are now wired back to extensions on her personal PBX. Anyone picking up a telephone in the house hears the dial tone provided by the personal PBX and is able to call between rooms, or place external outgoing calls if a trunk is available (and if Debbie configures the switch to allow external calls from the station in question). Incoming calls on the residential line ring on the kitchen phone, but they can be picked up by any extension using *group pickup call*. The other trunk is used for incoming business calls that ring only on her extension. If any call is unanswered, forwarding rules determine that the call is redirected to the voice mail/fax extension. The personal PBX is integrated with her home computer using a serial connection that supports isochronous communication. This allows her computer to both control the PBX and access audio media stream data associated with her extension. She also has her computer's voice/fax modem connected to its own dedicated extension. This extension is used for both voice mail and fax reception. Any time that an incoming fax call is detected on either trunk, it is routed to the voice mail/fax extension.

Figure 9-7. Personal telephone system solution scenario configuration

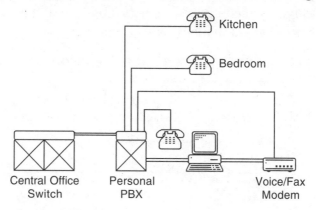

9.4.2. CTI Software Components

The software components that complete Debbie's CTI solution are shown in Figure 9-8. Debbie's mission-critical applications are the databases that track her clients and do her time and billing. Her client database is tied to her screen-based telephone application through a high-level CTI interface, which allows it to place calls and also to present client information pertinent to incoming calls using callerID. Two programmed telephony applications run on the computer in the background at all times. One is a call-logging application that observes and logs everything that takes place in the switch. The time and billing application monitors the logging application in order to identify calls, placed from anywhere in the house, that can be billed to clients. The other programmed telephony application is a messaging program that handles incoming faxes and voice mails on the fax line. It operates completely independently of the other CTI products, using a different CTI service boundary and switching domain. The last key feature of Debbie's solution is that the particular CTI client implementation she has installed supports a speaker phone feature, configured to be activated in conjunction with her extension. This means it is able to use the computer's speakers and microphone as a speaker phone, extending the capabilities of the actual (POTS) physical device element. Debbie's screen-based telephone application perceives a physical device element that has two auditory apparatuses.

Figure 9-8. Personal telephone system solution scenario software

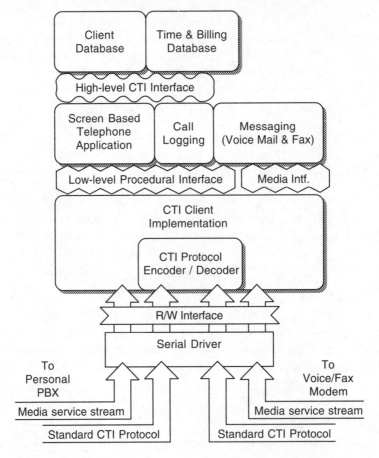

9.5. Personal Telephone Agent

Edmund, a photographer, set up a CTI solution for himself that involved customizing a personal telephone agent to handle calls on his behalf. Edmund's scenario is summarized in Table 9-5. (Refer to Chapter 2, section 2.7 for the complete scenario description.)

Table 9-5. Personal telephone agent solution scenario

Name:	Edmund
Occupation:	Photographer
Location:	Small business office
Switch:	Central office
Telephone station:	ISDN with analog terminal adapter
Type of computer:	Desktop computer
Type of CTI Interface:	Direct-connect
Applications used:	Personal agent software
	Screen-based telephone
	Fax software
Implemented by:	Third-party developer
Customization:	Prompts and logic customized
Future Plans:	Speech recognition
Benefits:	Edge on the competition
	Establishes more professional image
	Saves money
	Remote access capability

9.5.1. CTI System Configuration

The CTI system configuration for Edmund's CTI solution is shown in Figure 9-9. It involves an ISDN subscriber line to the central office switch, connected to a telephone station peripheral (a terminal adapter in this case), which in turn is connected to both his desktop computer and his old analog POTS telephone. The ISDN line corresponds to two different logical devices in the central office switch. One corresponds

519

to Edmund's telephone and one corresponds to Edmund's new personal agent. Forwarding rules are established such that if the one associated with the telephone is not answered, or if *do not disturb* is set, calls are forwarded to the agent's logical device. The speaker output from his computer is extended to his studio so that sounds made by his computer can be heard in the studio and darkroom.

Figure 9-9. Personal telephone agent solution scenario configuration

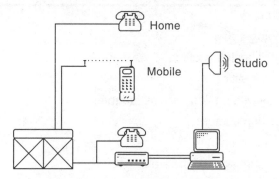

9.5.2. CTI Software Components

The software components that complete Edmund's CTI solution are shown in Figure 9-10. Edmund has installed a screen-based telephone application that monitors and controls the logical device associated with his telephone number. He uses this application for making and managing calls when he is in his office. He also can use it to invoke the *set do not disturb* service. In addition, he has installed a personal agent application that observes and controls the second logical device associated with the ISDN line. The personal agent has full access to media streams, so it can interact with callers, send faxes, take messages, etc. As described in the scenario in Chapter 2, when the personal agent is taking an important call, it searches for Edmund by announcing the call on the studio speakers and by calling his cellular phone and his home telephone. In each case it uses text-to-speech technology to speak aloud the callerID and other call associated information. It also records the caller's answers to certain questions and plays these recordings back.

Figure 9-10. Personal telephone agent solution scenario software

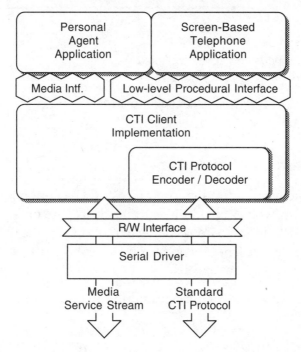

9.6. Interactive Voice Response System

Frances, the school vice-principal, set up a CTI solution to provide information to parents using a programmed telephony application for interactive voice response. Frances's scenario is summarized in Table 9-6. (Refer to Chapter 2, section 2.8 for the complete scenario description.)

Table 9-6. Interactive voice response solution scenario

Name:	Frances
Occupation:	Vice-principal
Location:	School
Switch:	Central office
Telephone station:	Analog POTS
Type of computer:	Desktop computer

Table 9-6. Interactive voice response solution scenario (Continued)

Type of CTI Interface:	Direct-connect
Applications used:	CTI application generator
Implemented by:	Frances and school whiz kid
Customization:	Daily updates by teachers
Future Plans:	More phone lines and Internet access
Benefits:	Better community service

9.6.1. CTI System Configuration

The CTI system configuration for Frances's CTI solution is shown in Figure 9-11. It involves an add-in card inside a spare personal computer. The add-in board is connected to a POTS line from the central office switch. There is no telephone set because locally there is no human interaction with calls.

Figure 9-11. Interactive voice response solution scenario configuration

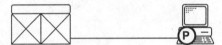

9.6.2. CTI Software Components

The software components that complete Frances's CTI solution are shown in Figure 9-12. Like the system configuration for this scenario, it is quite simple. A CTI application generator was used to develop a special programmed telephony application for the school information system. Because the hardware component in this case is an add-in board, a mapper is required in order to access the standard CTI protocols.

Figure 9-12. Interactive voice response solution scenario software

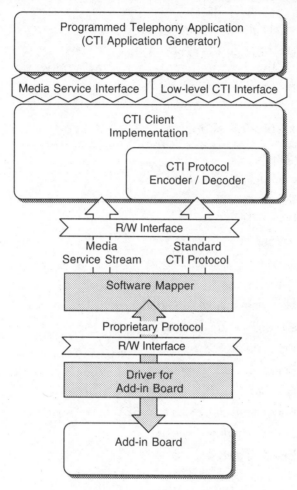

9.7. Help Desk

Gunther, the help desk technician, has constructed an extensive CTI solution consisting of a number of different server processes running on a computer that is shared across a local area network. Gunther's scenario is summarized in Table 9-7. (Refer to Chapter 2, section 2.9 for the complete scenario description.)

Table 9-7. Help desk solution scenario

Name:	Gunther
Occupation:	Help desk technician
Location:	University
Switch:	PBX
Telephone station:	Various
Type of computer:	Various
Type of CTI Interface:	Client-server
Applications used:	Multi-user call tracking database
	CTI application generators (two)
	Web server
	Screen-based telephones
Implemented by:	Gunther
Customization:	Each technician picks own SBT
Future Plans:	Screen sharing
Benefits:	Higher-quality service
	More timely service
	Cost savings

9.7.1. CTI System Configuration

The CTI system configuration for Gunther's help desk CTI solution is shown in Figure 9-13. Everything in the system is connected together using a LAN, and the university's PBX is used to deliver and queue telephone calls from help desk clients to technicians. A hybrid media/ CTI server is connected to the CTI interface of the PBX using a serial communication path for establishing a CTI stream. This hybrid server provides a fan-out of CTI call control information to authenticated clients, and also allows those clients to access media streams associated with any call by conferencing one of the lines connected to the built-in media server resources into the specified call. The media server that Gunther has installed is a very simple one that is capable only of playing prerecorded messages. The hub of Gunther's solution is the shared computer used to run the call tracking database, a call routing application, a voice processing application, and a Web server.

Figure 9-13. Help desk solution scenario configuration

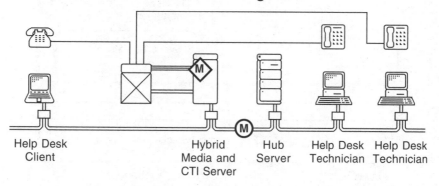

9.7.2. CTI Software Components

The software components running on the hub machine in Gunther's CTI solution are shown in Figure 9-14. These include the following:

- Multi-user call tracking database
  The multi-user database is accessed by all the other applications on this system and by the database clients running on the technician workstations. The database's client-sever protocol (not shown) also travels over the TCP/IP connection.

- Voice Processing
  The voice processing application is a programmed telephony application responsible for interacting with callers to the help desk number. This application uses the CTI interface to

Figure 9-14. Help desk solution scenario hub software

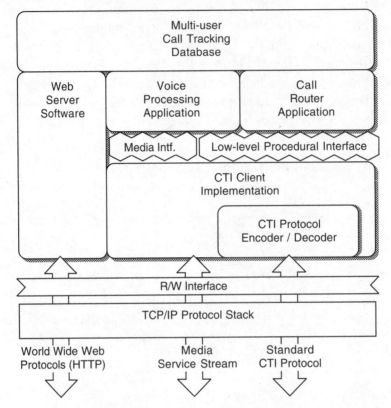

monitor for each new incoming call and, after binding a media service instance to it for playing prerecorded messages, questions each caller about the nature of the problem. The DTMF tones that are detected from the user are translated into the appropriate information to place in fields in the call tracking database. Once the new call has been appropriately classified, it is queued and the routing application takes over.

- Call router

  The call router is a programmed telephony application that uses the CTI interface to observe the calls queued for technicians, and to control routing them to the appropriate technician based on information in the call tracking database.

At the appropriate time, this application also establishes calls to those (like the user of the notebook computer in Figure 9-13) who request a position in the queue via a World Wide Web browser.

- Web server

    The Web server provides an alternate mechanism for joining the queue by directly creating an entry in the call tracking database.

Figure 9-15. Help desk solution scenario client software

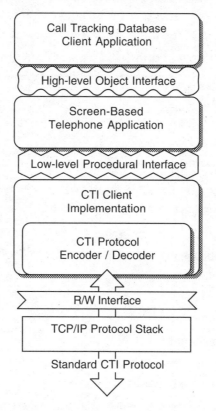

The software components running on the technician workstations are shown in Figure 9-15. In contrast to the hub machine, the software configuration of the technician workstations is quite simple. Each has an SBT of the technician's own choice and a database client application that accesses the multi-user call tracking database. (The database's client-sever protocol also travels over the TCP/IP connection, but is not shown here.) When a call is delivered to a particular technician, the appropriate call tracking information is also presented automatically.

9.8. Call Center

Henrietta, the manager of a successful virtual travel agency, has built her entire business around CTI technology that allows her travel agents to work from their own homes as part of a distributed call center. Henrietta's scenario is summarized in Table 9-8. (Refer to Chapter 2, section 2.10 for the complete scenario description.)

Table 9-8. Call center solution scenario

Name:	Henrietta
Occupation:	Manager of travel agency
Location:	Distributed
Switch:	Central office
Telephone station:	Centrex ISDN, Centrex analog
Type of computer:	Desktop and server
Type of CTI Interface:	Client-server
Applications used:	Multi-user client database
	CTI application generator
	Screen-based telephones
Implemented by:	Henrietta
Customization:	N/A
Future Plans:	Screen sharing
Benefits:	Makes the business model possible
	Improves customer satisfaction
	Boosts professional image of organization
	Improves employee morale
	Reduces call times and eliminates errors
	Minimizes business overhead

9.8.1. CTI System Configuration

The CTI system configuration for Henrietta's CTI solution is shown in Figure 9-16. Components exist in three distinct locations: the central office, Henrietta's basement, and the travel agent's home.

- The central office

  In addition to the central office switch, the central office houses a CTI server that provides access to a switching domain consisting of all of Henrietta's lines. Henrietta subscribes to Centrex and has both ISDN and analog lines. The analog lines are used for media services (voice processing and sending pager messages). ISDN is used everywhere else.

- Henrietta's basement

  The server that runs the distributed call center is in Henrietta's basement. A single machine is responsible for running everything.[9-1] It is a hybrid media server and CTI server, and simultaneously supports a CTI client implementation along with other software needed by Henrietta's business. ISDN is used to establish a dial-up communication path to the CTI server at the central office. Henrietta's basement also contains a dial-in bridge for her LAN. This supports dial-in access to the server for all of her agents.

- The travel agent's home

  Each travel agent has an ISDN line and a desktop computer. One ISDN B channel is used to establish a communication path with Henrietta's LAN in order to connect to the server. The other is used for voice and is accessed using an analog phone through a terminal adapter. (Note that the configuration diagram reflects the logical relationships, not the physical connections, so the communication path is

---

9-1. UPS — Yes, Henrietta does have a UPS (Uninterruptible Power Supply) protecting this server.

simply shown between the agent machine and the dial-in bridge, even though it physically passes through the central office switch.)

Figure 9-16. Call center solution scenario configuration

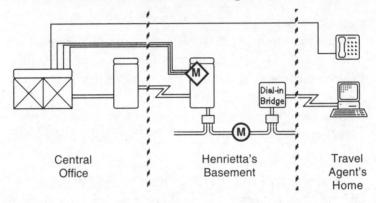

Central Office   Henrietta's Basement   Travel Agent's Home

9.8.2. CTI Software Components

Most of the software components that complete Henrietta's CTI solution are found on her central server as shown in Figure 9-17. These software components include:

- Hybrid media server and CTI server
  The hybrid media server and CTI server, is a single software component. It is a CTI Plug & Play CTI server implementation that has added value in the form of full-function media access to several analog lines through add-in media access boards installed in the server machine. The media services supported include sound record, sound playback, and modem.

- Auto-attendant application
  An auto-attendant application greets callers, identifies an appropriate agent, redirects calls, takes messages, and sends pages as necessary. The auto-attendant is a client of the CTI server running on the same machine.

530

- Multi-user client database

A multi-user client database tracks all information about the travel agency's clients and their travel plans. In the event that a given travel agent cannot be reached, another can use this database to support a given customer. The database's client-server protocol (not shown) also travels over the TCP/IP connection.

Figure 9-17. Call center solution scenario server software

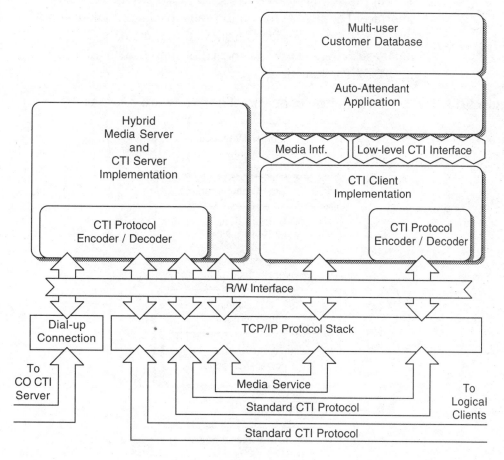

The software components running on the individual travel agents' systems are shown in Figure 9-18. These include:

- Screen-based telephone

    The SBTs used on the travel agent's machines have a unique feature that allows them to automatically navigate through the IVR menus associated with frequently called numbers.

- Client database application

    This application is linked to the SBT through the high-level interface. These allow incoming calls to trigger lookups and outgoing calls to be placed from within the database. The database's client-server protocol (not shown) also travels over the TCP/IP connection.

Figure 9-18. Call center solution scenario client software

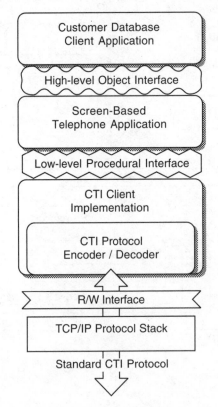

9.9. Conclusion

Regardless of the type or size of your organization, CTI will be an important tool for you in the future. Whether your goals are better service, cost savings, increased revenue, improved morale, or simply the ability to be more productive and effective, CTI can help you achieve these goals.

Having read this book, you should have sufficient understanding of the terminology, concepts, and technologies of the telephony and CTI industries to begin developing plans for CTI solutions of your own. This book should continue to serve as a resource as you plan for the implementations of CTI solutions or CTI components in your environment, and as you attempt to interpret product and service claims from vendors throughout the CTI value chain.

534

Bibliography

There are five reference documents that developers of CTI products may wish to consult:

- Apple Computer, Inc. *Telephone Manager Developer's Guide*, Mac OS SDK. (Cupertino, Calif.: Apple Computer, Inc., 1991).

- Apple Computer, Inc. *Telephony Apple Event Suite*, Mac OS SDK. (Cupertino, Calif.: Apple Computer, Inc., 1991).

- Microsoft Corporation. *Telephony Application Programmer's Interface (TAPI) Reference*, Microsoft Developer Network Library. (Redmond, Wash.: Microsoft Corporation, 1995).

- Microsoft Corporation. *Telephony Service Provider Programmer's Reference*, Microsoft Developer Network Library. (Redmond, Wash.: Microsoft Corporation, 1995).

- Versit. *Versit Computer Telephony Integration (CTI) Encyclopedia, Vols. 1–6*. (Versit, 1996).

These documents are complete reference documents intended for different audiences as described below.

Bibliography

Telephony Equipment Vendors and Network System Providers

Anyone developing telephony equipment or hardware products that connect to telephone lines should consult the *Versit CTI Encyclopedia* specifications. In particular, the following volumes should be used as reference material in developing the specifications for any CTI products:
- Volume 3: Telephony Feature Set
- Volume 4: Call Flow Scenarios
- Volume 5: Protocols

Hardware Vendors and Operating System Vendors

Those developing CTI client implementations or CTI server implementations should consult these volumes of the *Versit CTI Encyclopedia*:
- Volume 3: Telephony Feature Set
- Volume 4: Call Flow Scenarios
- Volume 5: Protocols
- Volume 6: Versit TSAPI

If the implementation is Mac OS based, consult the *Telephone Manager Developer's Guide*.

If the implementation is Windows-based, consult the Microsoft *Telephony Service Provider Programmer's Reference*.

Mac OS Application Developers

Those developing telephony-specific software should consult the following volumes of the *Versit CTI Encyclopedia*:
- Volume 3: Telephony Feature Set
- Volume 4: Call Flow Scenarios

Developers planning to use the Versit TSAPI programmatic interface should refer to:
- Volume 6: Versit TSAPI of the *Versit CTI Encyclopedia*.

Developers planning to use the Telephone Manager programmatic interface should refer to:

- the Apple *Telephone Manager Developer's Guide.*

Those developing mainstream applications that wish to be telephony-aware should consult:

- the *Telephony Apple Events Suite*

Windows Application Developers

Those developing telephony-specific software should consult the following volumes of the *Versit CTI Encyclopedia*:

- Volume 3: Telephony Feature Set
- Volume 4: Call Flow Scenarios

Developers planning to use the Versit TSAPI programmatic interface should refer to:

- Volume 6: Versit TSAPI of the *Versit CTI Encyclopedia.*

Developers planning to use the TAPI programmatic interface should refer to:

- the Microsoft *Telephony Application Programmer's Interface (TAPI) Reference*

Multiplatform Application Developers

Those developing telephony-specific software to run on multiple platforms or platforms other than Mac OS and Windows should consult the following volumes of the *Versit CTI Encyclopedia*:

- Volume 3: Telephony Feature Set
- Volume 4: Call Flow Scenarios
- Volume 6: Versit TSAPI

Bibliography

Index

Index

Index

Index

Index

About the Author

Michael Bayer is the president of Computer Telephony Solutions, a consulting firm specializing in the design and implementation of CTI technologies and products. His work in this field dates back over a decade—before the term "CTI" even emerged—and he is recognized as one of the pioneers of the CTI industry. Many important CTI initiatives can be traced to his groundbreaking technical work, extensive promotion for his vision of CTI, and participation in various standards-setting efforts Mr. Bayer has a technical background in computer-mediated communications and human interface design, and his technical credits include developing the Macintosh Telephony Architecture (MTA) and, more recently, playing a leading role in the development of the *Versit CTI Encyclopedia*.